Irina Polovodova Asteman

Benthic foraminifera and environmental change

Irina Polovodova Asteman

Benthic foraminifera and environmental change

A case study from the south-western Baltic Sea

Südwestdeutscher Verlag für Hochschulschriften

Impressum/Imprint (nur für Deutschland/only for Germany)
Bibliografische Information der Deutschen Nationalbibliothek: Die Deutsche Nationalbibliothek verzeichnet diese Publikation in der Deutschen Nationalbibliografie; detaillierte bibliografische Daten sind im Internet über http://dnb.d-nb.de abrufbar.

Alle in diesem Buch genannten Marken und Produktnamen unterliegen warenzeichen-, marken- oder patentrechtlichem Schutz bzw. sind Warenzeichen oder eingetragene Warenzeichen der jeweiligen Inhaber. Die Wiedergabe von Marken, Produktnamen, Gebrauchsnamen, Handelsnamen, Warenbezeichnungen u.s.w. in diesem Werk berechtigt auch ohne besondere Kennzeichnung nicht zu der Annahme, dass solche Namen im Sinne der Warenzeichen- und Markenschutzgesetzgebung als frei zu betrachten wären und daher von jedermann benutzt werden dürften.

Coverbild: www.ingimage.com

Verlag: Südwestdeutscher Verlag für Hochschulschriften GmbH & Co. KG
Heinrich-Böcking-Str. 6-8, 66121 Saarbrücken, Deutschland
Telefon +49 681 37 20 271-1, Telefax +49 681 37 20 271-0
Email: info@svh-verlag.de

Approved by: Kiel, CAU, Diss., 2008

Herstellung in Deutschland:
Schaltungsdienst Lange o.H.G., Berlin
Books on Demand GmbH, Norderstedt
Reha GmbH, Saarbrücken
Amazon Distribution GmbH, Leipzig
ISBN: 978-3-8381-3175-7

Imprint (only for USA, GB)
Bibliographic information published by the Deutsche Nationalbibliothek: The Deutsche Nationalbibliothek lists this publication in the Deutsche Nationalbibliografie; detailed bibliographic data are available in the Internet at http://dnb.d-nb.de.

Any brand names and product names mentioned in this book are subject to trademark, brand or patent protection and are trademarks or registered trademarks of their respective holders. The use of brand names, product names, common names, trade names, product descriptions etc. even without a particular marking in this works is in no way to be construed to mean that such names may be regarded as unrestricted in respect of trademark and brand protection legislation and could thus be used by anyone.

Cover image: www.ingimage.com

Publisher: Südwestdeutscher Verlag für Hochschulschriften GmbH & Co. KG
Heinrich-Böcking-Str. 6-8, 66121 Saarbrücken, Germany
Phone +49 681 37 20 271-1, Fax +49 681 37 20 271-0
Email: info@svh-verlag.de

Printed in the U.S.A.
Printed in the U.K. by (see last page)
ISBN: 978-3-8381-3175-7

Copyright © 2012 by the author and Südwestdeutscher Verlag für Hochschulschriften GmbH & Co. KG and licensors
All rights reserved. Saarbrücken 2012

Benthic foraminifera and environmental change

A case study from the south-western Baltic Sea

Dissertation

zur Erlangung des Doktorgrades
der Mathematisch-Naturwissenschaftlichen Fakultät
der Christian-Albrechts-Universität zu Kiel

vorgelegt von
Irina Polovodova

Kiel, 2008

Contents

Summary/Zusammenfassung	vii
Acknowledgements	xi
Introduction	1
CHAPTER 1 **Foraminiferal response to environmental change in Kiel Fjord (SW Baltic Sea)**	5
ABSTRACT	5
1.1 INTRODUCTION	5
1.2 STUDY AREA	7
1.2.1 Previous pollution surveys	7
1.3 MATERIAL AND METHODS	9
1.3.1 Sampling	9
1.3.2 Hydrographical measurements	9
1.3.3 Geochemical analysis	10
1.3.4 Foraminiferal studies	10
1.4 RESULTS AND DISCUSSION	11
1.4.1 Hydrography	11
1.4.2 Organic carbon and C/N ratio	11
1.4.3 Biogenic silica	12
1.4.4 Chlorine and phaeopigments	12
1.4.5 Trace metals pollution	12
1.4.6 Foraminiferal population density and species composition	15
1.4.7 Re-examination of Lutze's material	18
1.4.8 Invasion and opportunistic behaviour of *Ammonia beccarii*	18
1.4.9 Disappearance of *Ammotium cassis*	19
1.5 CONCLUSIONS	20
CHAPTER 2 **Recent benthic foraminifera from Flensburg Fjord: distribution and response to environmental change**	22
ABSTRACT	22
2.1 INTRODUCTION	22

2.2 STUDY AREA		24
2.3 MATERIAL AND METHODS		25
2.4 RESULTS		27
	2.4.1 General trends in foraminiferal distribution	27
	2.4.2 Cluster analysis	30
2.5 DISCUSSION		31
	2.5.1 Food supply	31
	2.5.2 Reliability of foraminiferal abundances	33
	2.5.3 Predominance of *Elphidium incertum* in the inner Flensburg Fjord	34
	2.5.4 Evidence of oxygen deficiency in sediments of Gelting Bay?	35
	2.5.5 Comparison with previous studies	35
	2.5.6. Changes in species composition	37
	2.5.7 Absence of *Eggereloides scaber*	38
2.6 CONCLUSIONS		38
CHAPTER 3	**Test abnormalities of recent benthic foraminifera in the western Baltic Sea**	40
	ABSTRACT	40
	3.1 INTRODUCTION	41
	3.2 REGIONAL SETTING	42
	3.3 MATERIAL AND METHODS	44
	3.4 RESULTS	46
		3.4.1 Hydrography	46
		3.4.2 Definition of abnormality modes	47
		3.4.3 Distribution and diversity of abnormal tests in Kiel Fjord	48
		3.4.4 Distribution and diversity of abnormal tests in Flensburg Fjord	49
		3.4.5 Distribution of abnormalities between living and dead assemblages	51
		3.4.6 Test abnormalities and sediment geochemistry	52
	3.5 DISCUSSION	53
		3.5.1 Abnormal tests as indicators of heavy metal pollution?	53
		3.5.2 Predominance of small test abnormalities	57
		3.5.3 Morphological constraints for the development of abnormal tests	57

3.5.4 Aberrant tests of *Ammonia beccarii* in Gelting Bay	57
3.5.5 Abnormalities as a result of abrupt salinity changes	60
2.6 CONCLUSIONS	61
Outlook	63
General conclusions	66
References	69
APPENDIX	I
APPENDIX 1. Faunal Reference List	II
APPENDIX 2. Kiel Fjord	III
Appendix 2.1 Station list	III
Appendix 2.2 Census data	V
Appendix 2.3 Living/Dead abundances from Lutze (1965) samples	VIII
Appendix 2.4 Correlation matrix of environmental and foraminiferal data	IX
APPENDIX 3. Flensburg Fjord	X
Appendix 3.1 Exon's census data	X
Appendix 3.2 Census data of this study	XI
APPENDIX 4. Test abnormalities	XII
Appendix 4.1 Absolute abundances of abnormality modes (Kiel Fjord)	XII
Appendix 4.2 Absolute abundances of abnormality modes (Flensburg Fjord)	XIII
Appendix 4.3 Correlations between abnormality modes and environmental data	XIV
APPENDIX 5 PLATES	XV

Summary

The Baltic Sea is a vulnerable ecosystem. It undergoes a high environmental variability due to the occasional ventilation of deep water by highly saline Kattegat water, which enters the Baltic through the narrow and shallow Danish Straits. During the past two decades, the frequency of major inflows decreased significantly, leading to extended periods of stagnation with reduced deep-water oxygenation. At the same time, a growing economy led to significant anthropogenic pollution in this area. The anthropogenic influx of different pollutants caused eutrophication, oxygen depletion and left elevated levels of trace metals in the sediments. These environmental changes are seen today at all trophic levels of living organisms and affected even one of the smallest inhabitants of the Baltic Sea: benthic foraminifera. In this thesis, foraminifera are evaluated as proxies of recent environmental change from both natural and anthropogenic origin in two shallow fjords of the Kiel Bight (SW Baltic Sea).

Kiel Fjord is moderately polluted by trace metals, and this fjord was chosen as the study area for tracing the anthropogenic influence on benthic foraminifera. Analysis of foraminiferal population density showed a patchy distribution, which mainly reflects food availability. Significant changes in species composition were observed in 2005-2006, as compared to the 1960s. The unfavourable salinity conditions and the absence of a deep halocline in Kiel Fjord have caused the disappearance of the arenaceous species *Ammotium cassis*. A sediment core from the outer Kiel Fjord was studied to trace the occurrence of *Ammotium cassis* and to investigate the history of environmental changes in this area over the last century. On the other hand, *Ammonia beccarii*, a eurihaline and highly opportunistic calcareous species, which tolerates elevated levels of nutrients and trace metals, has recently invaded Kiel Fjord. Increased test abnormalities in *A. beccarii* correlated to high trace metal levels during springtime. This relationship mirrored the sensitivity of juveniles to environmental stress after reproduction of benthic foraminifera.

Foraminiferal assemblages of entire Flensburg Fjord were surveyed for the first time. Five foraminiferal biofacies were distinguished and were mainly controlled by food availability. Comparison with previous studies, conducted in the outer fjord in the late 1940s and 1970s, revealed a decline of *Ammotium cassis* in the outer fjord, the flourishing of *Ammonia beccarii* in the Gelting Bay and the dominance of *Elphidium incertum* in the inner fjord. These changes are similar to those reported recently from other fjords of the SW Baltic Sea, and are associated with the generally lower intensity and frequency of major Baltic inflows since the 1960s.

The distribution of foraminiferal test abnormalities was studied in both fjords. Eighteen modes

Summary

of abnormal tests were recognized. According to morphological criteria, the modes were classified into 5 groups: chamber-, aperture-, umbilicum-, coiling- and test abnormalities. The highest abnormality frequencies were observed in the outer parts of the fjords facing salt-water inflows. In the inner fjords elevated levels of heavy metals apparently led to higher percentages of abnormal tests in places. The validity of using abnormal foraminiferal tests as indicators of anthropogenic pollution is discussed. I developed a conceptual model depicting the relationship between salinity tolerance of certain foraminiferal species and the development of test abnormalities. This model indicates that test abnormalities cannot be used exclusively as pollution indicators.

Zusammenfassung

Die Ostsee ist ein besonders anfälliges Ökosystem auf Grund hoher Umweltvariabilitäten, die mit dem Einstrom von salzreichem Bodenwasser aus dem Kattegatt einhergehen. Im Laufe der letzten zwei Jahrzehnte nahmen die Einstromereignisse deutlich ab, was zu Stagnation mit periodischer Sauerstoffabnahme im Bodenwasser führte. Bedeutende Umwelteinflüsse durch die starke Wirtschaftsentwicklung führten zu Eutrophierung, Sauerstoffmangel und Schwermetallbelastung in Bodensedimenten. Diese Umweltveränderungen zeigen sich heuzutage in allen Nahrungsebenen und beeinflussen sogar einen der kleinsten Bewohner der Ostsee, die benthische Foraminiferen. Die vorliegende Arbeit hat zum Ziel, die Foraminiferen als Proxies der rezenten Umweltsveränderungen, natürliche und anthropogene, in zwei flachen Förden der Kieler Bucht (SW Ostsee) auszuwerten.

Auf Grund einer mittelmäßigen Schwermetallverschmutzung wurde die Kieler Förde als Untersuchungsgebiet ausgesucht, in dem man den anthropogenen Einfluss auf die benthischen Foraminiferen verfolgen kann. Die Populationsdichte der Foraminiferen zeigte eine ungleichmäßige Verteilung und eine empfindliche Reaktion auf Nahrungszufuhr. Ein Vergleich mit historischen Daten aus den vierziger und sechziger Jahren zeigte deutliche Veränderungen in der Artenzusammensetzung und der Populationsdichte. *Ammonia beccarii*, eurihalin und eine sehr opportunistische, kalkschalige Art mit hoher Toleranz gegenüber erhöhten Nährstoff- und Schwermetallkonzentrationen, ist in die Kieler Förde eingedrungen. Der Sandschaler *Ammotium cassis* ist hingegen auf Grund eines niedrigeren Salzgehalts und fehlender Sprungschicht in der Kieler Förde seit den neunziger Jahren verschwunden. Ein Sedimentkern der Kieler Außenförde wurde untersucht, um die historischen Vorkommen von *Ammotium cassis* in Kieler Förde während der letzten hundert Jahre zu beschreiben und die Geschichte der Umweltveränderungen in diesem Gebiet zu verfolgen. Die erhöhten Schalenmissbildungen von *Ammonia beccarii* korrelieren mit hohen Schwermetallkonzentrationen im Frühjahr. Dies spiegelt sich in der Fortpflanzung von Foraminiferen und in der besonderen Empfindlichkeit der Juvenilen gegenüber umweltbedingtem Stress wider.

Die Benthosforaminiferen-Gemeinschaften in der Flensburger Förde wurden zum ersten Mal untersucht und reflektierten eine hohe natürliche Variabilität. Die Untersuchungen lassen fünf unterschiedliche Foraminiferen-Biofazies erkennen, die hauptsächlich durch Nahrungszufuhr gesteuert werden. Ein Vergleich mit früheren Daten aus den 40er und 70er Jahren zeigte eine Abnahme von *Ammotium cassis* in der Flensburger Außenförde eine Acme von *Ammonia beccarii*

Zusammenfassung

in der Geltinger Bucht und die Dominanz der endobenthischen Art *Elphidium incertum* in der Innenförde. Diese Veränderungen sind ähnlich im Vergleich zu jenen, die in anderen Förden der Kieler Bucht beobachtet wurden und mit der allgemein verringerten Intensität und Häufigkeitsabnahme der Einstromereignisse seit den sechziger Jahren in Verbindung gebracht wurden.

Weiterhin wurde die Verteilung von Foraminiferen-Schalenmissbildungen in beiden Förden untersucht. Dabei wurden achtzehn Missbildungstypen identifiziert. Aufgrund von morphologischen Kriterien wurden alle Typen in fünf Gruppen klassifiziert: die Kammer-, Apertur-, Nabel-, Windungs- und Schalenmissbildungen. Die höchste Anzahl von Missbildungen wurde in den äußeren Teilen beider Förden beobachtet, welche dem Salzwasser-Einstrom zugewandt sind. Dagegen sind in den Innenförden die erhöhten Schwermetallbelastungen für das häufige Missbildungsvorkommen verantwortlich. Im Folgenden wird die Stichhaltigkeit der Foraminiferen-Anwendung als Bioindikatoren durch anthropogene Umweltverschmutzung diskutiert. Diese Betrachtungsweise erlaubt die Entwicklung eines Modells, das die Verhältnisse zwischen Salinitätstoleranz und Entwicklung der Foraminiferenschalenmissbildungen beschreibt. Dieses Modell zeigt, dass man die Schalenmissbildungen als exklusive Verschmutzungsindikatoren nicht benutzen kann.

Acknowledgements

First of all I would like to thank my supervisors Prof. Dr. Wolf-Christian Dullo and Dr. Joachim Schönfeld for the initiation of this project and providing me with an excellent working place at IFM-GEOMAR. Prof. W.-Ch. Dullo supported my research with the participation of scientific conferences to present my results and gave me an opportunity to finish this project. I highly appreciate his support, helpful advices and very contagious optimism. Dr. Joachim Schönfeld provided me with a continuous support, helpful comments and valuable suggestions. In him I've found a very good scientific advisor and tutor, which was always standing by to answer my questions or to dispel my doubts.

Also I would like to express my sincere gratitude to the crew of RV Polarfuchs (Holger Meyer and Helmut Schramm), Dr. Volker Liebetrau and his diving buddy, Michael Albrecht, for the help with sampling.

I am especially thankful to Dr. Heide-Marie Kassens for giving me an opportunity to stay in May-June 2007 as a guest scientist at IFM-GEOMAR and for her wholehearted support during this study.

This study was funded by German Academical Exchange Service (DAAD), Research Grant of Otto-Schmidt Laboratory (AARI, SPb) and Leibniz Award DFG DU 129-33.

During my stay at St. Petersburg State University, several persons were of great help and I would like to thank all of them for their support. Prof. Dr. Gennady Belozerskiy was my supervisor, while staying at SPbSU. Dr. N. Sheynerman, Dr. M. Opekunova and Dr. S. Bonar provided me with microscoping facilities at the Laboratory of Environmental Monitoring (SPbSU). Prof. Dr. V.N. Movchan and Prof. Dr. V.V. Dmitriev gave me an opportunity to present my results at scientific conferences. Dr. Ekaterina Eldina and Dr. Irina Gembitskaya (St. Petersburg State Mining Institute) afforded the technical support with uncoated SEM imaging.

For the discovery of an incredible world of the FORAMINIFERA I would like to thank sincerely Prof. Dr. Tomas Cedhagen (University of Aarhus), Prof. Dr. Jan Pawlowski (University of Geneva), Prof. Dr. Gudmundur Gudmundsson (Nat. Hist. Museum, Reykjavik) and Prof. Dr. Elisabeth Alve (University of Oslo). From these four people I've learned the fascination of these tiny creatures. Elisabeth Alve is thanked also for the productive discussions and valuable advices during this study.

The discussions with Prof. Dr. F. Jorissen (University of Angers) made us to initiate the genetical identification of one of the world's most problematic *Ammonia* species. Great thanks to

Acknowledgements

Dr. Magali Schweizer (University of Edinburgh), who helped to solve this taxonomical problem by means of DNA analysis.

I am also very grateful to Prof. Dr. Wolfgang Kuhnt (University of Kiel), who gave me a continuous access to the archives of the Micropaleontological Museum and answered all my questions.

Junior-Prof. Dr. Vasil Golosnoy (Institute of Statistics and Econometrics, University of Kiel) and Dr. Eugenya Kandiano (IFM-GEOMAR) helped with statistical analysis. Their warm encouragement and help are highly appreciated.

I must acknowledge the colleagues, who have contributed towards the pleasant and friendly working environment at IFM-GEOMAR. Therefore, in no particular order I would like to thank Almuth Harbers, Dr. Syee Weldeab, Dr. Brian Haley, Roland Stumpf, Claudia Ehlert, Prof. Dr. Martin Frank, Dr. Nadja Kachro and Dr. Andres Rüggeberg, and everyone else from the Paleoceanography Group. Many thanks go to the technical stuff of the IFM-GEOMAR and Institute of Geosciences, and, namely, Jutta Heinze, Anna Kolevica, Lulzim Haxhiaj, Bettina Domeyer, Anke Bleyer, Stefan Sommer, Udo Laurer, Ute Schuldt, Ulrike Westernströer and Eduard Mezhyrov.

Anna Nikulina (IFM-GEOMAR), my classmate, my colleague and my best friend, I especially thank you for your endless support, taking a sober view of things all the time, and just for you being by my side during the last 3 years!

And of course this thesis would have never been finished without a continuous support of my family, which was always standing by, in spite of huge distances. I wish to thank my mother Anna, my father Valentin and my precious brothers Igor and Ivan for their loving support during all these years.

Introduction

Fjords of the Baltic Sea are important transport, filter and buffer systems, because they promote the absorption and accumulation of different organic and inorganic compounds brought from the land (Schiewer & Gocke, 1995). At the same time, the fjords are often highly urbanized areas, which undergo a human-induced impact and represent exemplary research areas to study faunal responses to environmental changes.

Benthic foraminifera, hard-shelled protozoans, are a suitable tool for tracing such kind of response. They inhabit the bottom water-sediment interface or dwell in the sediments, which serve as natural accumulation reservoirs for various chemical substances. Foraminifera are highly abundant and diverse, occur in all marine environments, and possess the ability to record environmental variations quite quickly via changes in species composition, population density, and the development of abnormal tests. Due to their high abundances and relatively small size of usually less than one millimetre, these protozoans provide a statistically significant sample size from a few cubic centimetres of sediment, which can be retrieved with minimal environmental impact. Then, the hard exoskeleton (test) of foraminifera preserves in the sediments after their death and provides a record of environmental variability through time. Thus, the faunal record is very valuable in environmental studies, in particular, where a reference or undisturbed site is missing. At the same time, the faunas help to answer the question «if changes of benthic community are human induced or represent the natural fluctuations?».

There are two basic approaches to utilize the benthic foraminifera as indicators of environmental change:
1) Examination of foraminiferal populations from surface sediment samples in order to assess the current state of the benthic ecosystem.
2) Study of faunal successions in sediment cores, determining the changes through time and giving the reference conditions at areas with a very long history of pollution.

The main objective of this dissertation is the evaluation of benthic foraminifera dwelling in the SW Baltic Sea as indicators of modern environmental change. For this purpose, the first of the above-mentioned approaches was used and namely the living (stained) benthic foraminifera from Kiel and Flensburg Fjords were examined together with hydrographical and geochemical parameters of bottom waters and surface sediments. The species distribution patterns and population density in Kiel Bight were compared to previous studies in order to find out how foraminiferal assemblages changed over the past decades. At the first stage of this study a

Introduction

particular attention was paid to anthropogenic changes, which have taken place in the Kiel Bight since the beginning of the last century.

Kiel Fjord was considered as a moderately polluted environment, subjected to the impact from shipbuilding industry, numerous military, sport and leisure harbours and intensive traffic through Kiel Canal. In order to reveal the response of benthic foraminifera to human-induced stress in Kiel Fjord, surface sediment samples were analysed for a set of trace metals: Cu, Sn, Zn and Pb, which are associated with the shipbuilding industry. Copper, tin and zinc have been used, as biocides in antifouling paints (Bellinger and Benham, 1978; Clark et al., 1988; Helland and Bakke, 2002), whereas lead is known to come from boat and ship exhaust systems (Abu-Hilal and Badran, 1990) and also forms the pigment basis of anticorrosives and primer paints (V.-Balogh, 1988).

However, it turned out that the natural variability is of greater importance for benthic foraminifera in Kiel Bight and changes in salinity associated with salt-rich bottom water inflows from Kattegat were the main controls. Indeed, as it is shown in **Chapter 1**, the arenaceous species *Ammotium cassis*, which was abundant in the 1960s and 1990s, has recently disappeared from Kiel Fjord. Long-lasting conditions of reduced salinity, which inhibited the formation of a stable halocline, that is necessary for nutrition of this foraminifer, are considered as the main reason for the decline of *A. cassis*. Along with their decline, Kiel Fjord was occupied by the calcareous *Ammonia beccarii*, a highly opportunistic and eurihaline species.

The distribution of living benthic foraminifera in Flensburg Fjord is discussed in **Chapter 2**. This fjord currently represents a resort area and faces the salt-water inflows at the most in Kiel Bight. My study presents the first description of foraminiferal fauna from the entire Flensburg Fjord. Five assemblages are described in this chapter. They are distributed mainly according to food availability. Gelting Bay is given here as an exception. It shows a high environmental variability, which is reflected in active hydrodynamics, sediment erosion, redeposition and low food availability. The high abundances of infaunal *E. incertum* at the sediment surface and enhanced porosity of *A. beccarii* tests were suggested as indirect evidences of seasonal oxygen depletion. To check this assumption, the *Ammonia/Elphidium* Index was used, as a proxy of hypoxia. In order to assess multidecadal changes, data of species composition from the outer Flensburg Fjord were compared to previous studies from the 1940s and 1970s.

Chapter 3 deals with morphological changes of foraminiferal tests reflected in the development of different test abnormalities, which are recently applied as indicators of anthropogenic impact. Distribution and diversity of test abnormality modes in both fjords are summarized. The validity of abnormal tests applied as proxies of human-induced impact in areas

Introduction

with high environmental variability is discussed by the example of Gelting Bay. Possible reasons for the shell loss, which was observed in Gelting Bay, are strained. On the basis of abnormality distribution patterns and the results of the literature data, a conceptual model is proposed, that shows the relationship between the development of abnormal tests and the salinity tolerance of *Ammonia* species.

The **Outlook** chapter gives an overview of changes in foraminiferal communities over multidecadal timescales. For this purpose, a short core, from the outer Kiel Fjord, was analysed. The record of environmental changes goes back to the middle of the 19th century. The history of anthropogenic impact, paleoceanography and foraminiferal dynamics are discussed.

All results of this thesis are summarized in **General Conclusions**.

Introduction

Chapter 1

Foraminiferal response to environmental changes in Kiel Fjord (SW Baltic Sea)

Published as:
Nikulina, A., Polovodova, I., and Schönfeld, J., 2007. Environmental response of living benthic foraminifera in Kiel Fjord, SW Baltic Sea, *eEarth Discuss.*, 2, 191–217. Available online at: www.electronic-earth-discuss.net/2/191/2007/

ABSTRACT

The living benthic foraminiferal assemblages in Kiel Fjord (SW Baltic Sea) were investigated in the years 2005 and 2006. The faunal studies were accomplished by geochemical analyses of surface sediments. In general, sediment pollution by copper, zinc, tin and lead is assessed as moderate in comparison with levels reported from other areas of the Baltic Sea. However, the inner Kiel fjord is still exposed to a high load of metals and organic matter due to enhanced accumulation of fine-grained sediments in conjunction with potential pollution sources as shipyards, harbours and intensive traffic. The results of our survey show that the dominant environmental forcing of benthic foraminifera is nutrients availability coupled with human impact. A comparison with faunal data from the 1960s reveals apparent changes in species composition and population densities. The stress-tolerant species *Ammonia beccarii* invaded Kiel Fjord. *Ammotium cassis* had disappeared that reflects apparently the changes in salinity over the last 10 years. These changes in foraminiferal community and a significant increase of test abnormalities indicate an intensified environmental stress since the 1960s.

1.1 INTRODUCTION

The previous studies in the Kiel Bight only gave a very short description of foraminiferal distribution, though they were started in 19th century (Möbius, 1888). Ecological observations of foraminifera were initiated by Rhumbler (1935), who used rather descriptive than quantitative methods of investigation. Next, Rottgardt (1952) distinguished three different foraminiferal assemblages in the Baltic Sea, which are distributed according to the salinity pattern: marine, brackish-marine (fjords and shallow areas of the Kiel Bight), and brackish faunas. A detailed taxonomical and ecological overview on benthic foraminifera in the south-western Baltic Sea was provided by Lutze (1965), who found out that temperature and salinity rather than substrate were the main ecological controls on foraminiferal distribution in this area. Vice versa, Wefer (1976) observed that the abundances of foraminifera in sediments off Bokniseck (open Kiel Bight) were regulated by substrate features, hydrodynamics and oxygen content of the bottom water. Foraminiferal food preferences in the open Kiel Bight were described by Schönfeld and Numberger (2007b), who reported two reproduction events of *Elphidium excavatum clavatum* following the spring bloom and suggested the "bloom-feeding" strategy of this species.

The benthic foraminiferal distribution in Kiel Fjord has been left out of sight, with the exception of 4 stations investigated by Lutze in 1962-1963, which were taken as reference points for our study. Over the 20th century, Kiel Fjord has experienced a strong anthropogenic impact. For monitoring purposes, the foraminiferal response to environmental changes attracts attention under the aspect of rising ecological problems.

A number of studies addressed the foraminiferal reactions to changing environmental parameters as salinity, temperature, oxygen, food availability, pH, (e.g. Bradshaw, 1957, 1961; Boltovskoy et al., 1991; Moodley and Hess, 1992; Alve and Murray 1999; Stouff et al., 1999ab; Gustafsson and Nordberg 2001; Le Cadre and Debenay, 2003), contamination by trace metals (Ellison et al., 1986; Sharifi et al., 1991; Alve, 1991; Alve and Olsgardt, 1999; Yanko et al., 1998; Debenay et al., 2001) and sewage effluents (e.g. Watkins, 1961; Schafer, 1973; Tomas et al., 2000). A decrease of population density, reproduction capability, enhanced mortality, and increasing frequency of test abnormalities were observed under the high trace metal or organic matter levels (Schafer, 1973; Samir & El Din, 2001; Bergin et al., 2006; Burone et al., 2006; Ernst et al., 2006; Di Leonardo et al., 2007). On the other hand, it was shown that population density of foraminifera may increase in vicinity of sewage outfalls (Watkins, 1961; Tomas et al., 2000). Culture experiments revealed that *A. beccarii* produces abnormal chambers at 10-20 µg/l of copper in seawater (Sharifi et al., 1991; Le Cadre & Debenay, 2006) and dies at concentrations exceeding 200 µg/l (Le Cadre & Debenay, 2006). Therefore, foraminifera appear to be a rather sensitive tool for the monitoring of pollution, though should be used with caution, because their distribution is determined by numerous environmental variables (Alve and Olsgardt; Stouff et al., 1999ab; 1999; Le Cadre & Debenay, 2006).

The aim of this study was (1) to describe the distribution of living (stained) benthic foraminifera in the Kiel Fjord, (2) to investigate the distribution pattern of main geochemical parameters of surface sediments, (3) to outline the level of pollution by trace metals, and (4) to assess the foraminiferal response to environmental changes during the past decades.

1.2 STUDY AREA

Kiel Fjord is a 9.5 km long, N-S extending and narrow inlet of south-western Kiel Bight (54°19' – 54°30'N; 10° 06' – 10°22'E). The Friedrichsort Sound divides the fjord into a southern, inner fjord with width to 250 m, and a northern, outer fjord, which expands up to 7.5 km and passes into Kiel Bight (Fig. 1.1). The inner Kiel Fjord is mostly 10 to 12 m deep. A system of up to 16 m deep channels connects the inner with the outer fjord. The outer fjord itself is more than 20 m deep.

As the entire Kiel Fjord is relatively shallow and isolated, its hydrographical characteristics weakly depend on the salt-rich inflow water from the Belt Sea. The only river discharging fresh water into Kiel Fjord is the Schwentine.

The water masses of the inner fjord are homogenously mixed, except during summer. Then, surface water has a temperature up to 16°C and a salinity of about 14 units. The underlying deep water has a temperature of about 12°C and a salinity of up to 21 units. In winter, the temperature does not change significantly with depth and may decrease to 2°C. The salinity is constant with depth as well (Schwarzer and Themann, 2003).

Coastal and near-shore erosion of Pleistocene till is the most important source of sediment in this area. Lag sediments with coarse sand and gravel prevail in the shallow coastal areas. They pass into sandy muds and silts in the deeper basins. In the innermost fjord, dark organic-rich muds are encountered even in shallow areas. Sand veneers are found in the Friedrichsort Sound due to relatively strong currents between inner and outer fjord (Schwarzer and Themann, 2003).

Kiel Fjord has seen a strong anthropogenic impact for the last 70 years by town infrastructure, shipyards, military and sport harbours and the intense traffic through Kiel Canal. The shipbuilding industry has led to a substantial trace metal pollution in places. Dredging to keep the seaways clear, and the ship traffic itself have caused a strong disturbance of surface sediments.

1.2.1 Previous pollution surveys

Despite the long-term anthropogenic load in study area, reports on the early history of pollution of Kiel Fjord are rare. Recently, the monitoring of metals concentration at a few stations in Kiel Bight by the Institute for Marine Research, Warnemünde (IOW) indicates no significant temporal trend in trace metal content for 1998-2000 with respect to the observed high interannual variability (e. g.

Nausch et al., 2003b, Pohl et al., 2005). Kiel Fjord itself is considered by LANU (The Regional Environmental Protection Agency of the Bundesland Schleswig-Holstein) as one of the most important local hot spots of cadmium, lead, copper, and zinc contamination in the coastal waters of Schleswig-Holstein. In the year 2000 for instance, the concentrations of Cu, Zn and Pb in sediment fraction <20 mm were 82, 300 and 130 mg/kg in the inner fjord correspondingly (Haarich et al., 2003), whereas in outer fjord Cu, Zn and Pb content was estimated to 30, 210 and 60 mg/kg respectively (LANU archive: Ostseemonitoring Programme).

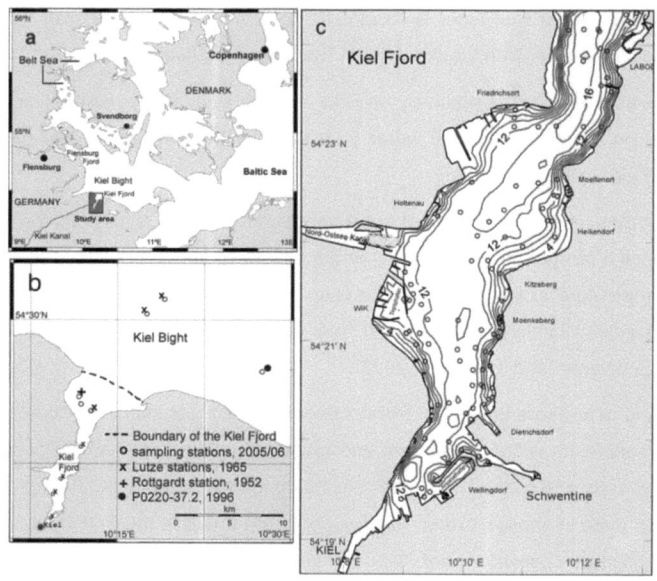

Figure 1.1: Study area: a – SW Baltic Sea, b – outer Kiel fjord, c – inner Kiel Fjord with bathymetry (m). Circles indicate here the sampling stations.

No clear temporal trend of metal concentrations in 1995-2004 was observed in sediments of Kiel Fjord. Extremely high concentrations of organically bound tin (407 – 2556 µg TBT-Sn/kg) were found in the fjord sediments; they are supposed to cause the aberrant changes in reproduction system of the periwinkle (LANU, 2001a). High concentrations of Cu and Zn were found in fish (Senosack, 1995) and mussels (ter Jung, 1992) from the inner Kiel Fjord. But the organisms in the outer fjord showed the lowest metals content for all Schleswig-Holstein waters.

Kiel Fjord has been affected by eutrophication induced by a high load of nutrient and organic carbon from the city and surrounding area (Gerlach, 1984). Herein, the nutrient concentrations and primary production showed a southward increase to the inner fjord (Schiewer and Gocke, 1995). The construction of a central treatment plant (Bülk, Klärwerk) in 1972 has reduced the input of nitrogen and phosphorus significantly (Kallmeyer, 1997, Rheinheimer, 1998), but the deep-water oxygenation improved not earlier than in the 1990s (Gerlach, 1996; Haarich et al., 2003; LANU, 2003). Nevertheless, oxygen deficiency may occur at specific weather conditions in the fjord regularly in late summer due to a stable water stratification (Gerlach, 1990).

1.3 MATERIAL AND METHODS

1.3.1 Sampling

This study is based on 89 surface sediment samples collected at 4.5-18.1 m water depth between December 2005 and May 2006 on seven daily cruises with R/V Polarfuchs. The samples were retrieved with a Rumohr corer with a plastic tube of 55 mm inner diameter and a Van-Veen Grab. The latter was used when sandy sediments were encountered. The Ruhmor corer was deployed three times at each station in order to avoid errors associated with spatial patchiness. The uppermost centimetre of the sediment was removed on each deployment with a spoon, and with cut-off syringes when a Van-Veen Grab was used. The sediment was placed into a glass vial, thoroughly mixed, and subsamples for geochemical analyses were taken from this mixture at first. The remaining sample was transferred to a PVC vial, and preserved and stained with a solution of 2 g Rose Bengal per litre ethanol in order to mark foraminifers living at the time of sampling (Murray and Bowser, 2000).

1.3.2 Hydrographical measurements

The salinity, temperature and dissolved oxygen content of the overlying water in the Rumohr corer tube was measured on board with Oxi- and Conductivity meters (WTW Oxi323/325 and LF320). As the measurements were made within minutes after retrieval, and air temperatures were not substantially higher than the water temperatures, we consider these values as representative for the near-bottom water. In the Schwentine river mouth, at three stations CTD-profiles were done with WTW Profiline 197 TS in 1-m intervals to locate the boundary between riverine fresh water and higher-saline fjord waters.

1.3.3 Geochemical analysis

Subsamples for geochemical analysis were freeze-dried and powdered in an agate mortar. Measurements of C_{org}, total carbon (TC) and total nitrogen (TN) were performed with a Carlo Erba NA-1500-CNS analyzer at IFM-GEOMAR with accuracy better than ±1.5%. Chlorophyll a and phaeopigments were determined after acetone extraction with a Turner TD-700 Fluorometer at IFM-GEOMAR. The precision of the method is ±10%. Biogenic silica (opal) measurements were done according to an automated leaching method for the analysis of SiO_2 in sediments and particulate matter described by Müller and Schneider (1993) using a Skalar 6000 photometer with precision ±1%. For trace metal analyses, the sediment samples were digested in a HNO_3-HF-$HClO_4$-HCl mixture solution. The solution was diluted and measurements were performed with an AGILENT 7500cs ICP-MS at the Institute of Geosciences, University of Kiel (Garbe-Schönberg, 1993). Blanks and the international standard MAG-1 were repeatedly analysed together with the samples in order to evaluate the precision and accuracy of the measurements. The accuracy of analytical results as estimated from replicate standard measurements was better than ±1.5%.

1.3.4 Foraminiferal studies

The sub-samples for foraminiferal analysis were stored in a fridge for two weeks to effect a sufficient staining with Rose Bengal. The samples were first passed through a 2000 µm screen in order to remove mollusc's shells or pebbles, and then gently washed through a 63-µm sieve. Sediments of the Baltic Sea have a high content of organic detritus. After drying, the detritus creates a film layer on the sample, which has to be disintegrated before picking (Lutze, 1965). In order to achieve a separation of the organic detritus, the 63—2000 µm size fraction was transferred into a cylinder with some tap water and left for a while. Then the supernatant water was poured through a filter paper to collect the suspended organic debris. During drying, the organic flocks stuck to the filter paper and foraminiferal tests could be easily brushed off (Lehmann and Röttger, 1997). The 63—2000 µm and >2000 µm fractions were dried at 60°C, weighed, and splitted. Well-stained foraminifers that were considered as living at the time of sampling were picked from respective aliquots, sorted at species level, mounted in Plummer cell slides and counted. Both normal and abnormal tests were counted separately. The standing stock was expressed as number of specimens per 10 cm^3 of sediment. The main species were photographed with Cam Scan Scanning Electronic Microscope at the Institute of Geosciences, Kiel University.

1.4 RESULTS AND DISCUSSION

1.4.1 Hydrography

The temperature and salinity of near-bottom water in Kiel Fjord showed a pronounced seasonality. Temperature decreased from 8°C on average in December 2005 to 2°C in February, and raised again to 7°C in May 2006. In December 2005, the near-bottom water showed the highest salinity with 23.2 units and minimum values of 16.5 units in May.

In the Schwentine river mouth, the boundary layer between riverine fresh water and saline fjord water was encountered at approximately 1 m depth in February. With an average discharge of 7.3 m^3/s (Schulz, 2000), the Schwentine substantially freshens the waters of the inner fjord.

The oxygen concentration mostly exceeded 400 µmol/l and decreased slightly only in the deep basins. The saturation levels varied from 58 % to 100 %. As such, a sincere oxygen deficiency in the near-bottom waters of Kiel Fjord was not recognized.

The oxygen content of near-surface sediments was measured with a Unisense microelectrode (Revsbech, 1989) in a short core taken from the inner fjord at the beginning of December 2005. The overlying water had oxygen saturation 71 %; the sediments were muddy-sand. At 1 mm sediment depth, the oxygen saturation was still more than 50 %, and a zero oxygen level was encountered at 3.5 mm. As compared with a usual 2 to 5 cm thick oxic layer in normal marine settings, the oxygenated surface layer in this core was quite thin.

1.4.2 Organic carbon and C/N ratio

The organic carbon content in the surface sediments ranged from 1% in Friedrichsort Sound to 7.8% in muddy sediments of the inner fjord (Fig. 1.2), and it is negatively correlated with the sand content (r=0.793, n=89). Though the changes in mean C_{org} values through the year were not substantial, we observed an increased C_{org} content associated with the spring bloom in February and March (Graf et al., 1982, Wasmund et al., 2005). Generally, the C_{org} content was higher than reported by Leipe et al. (1998) for the open Mecklenburg and Kiel Bights (5% for the fine fraction).

The mean C/N ratio depicts a substantial input of organic matter from the hinterland (Fig. 1.2). The C/N ratio increases southwards from 4 in the outer fjord to 15 in the inner fjord, which is in the range of values for the southern Baltic Sea (Pertillä et al., 2003). Seasonally, the C/N ratio changed not significantly but has the lower values in February-March that probably mirrors the accumulation of fresh detritus characterized by low C/N values of 5.6 to 7 (Graf et al., 1982).

1.4.3 Biogenic silica

Biogenic silica (opal) content in surface sediments of Kiel Fjord was higher in spring as compared to December (0.1 wt.% to 8 wt.%), and showed a maximum in the inner fjord (Fig. 1.2). The maximum of diatom biomass and biogenic silica flux to the sea floor was recorded in early April in the SW Baltic Sea (Wasmund et al., 2005, 2006). Apparently, the increase of opal in sediments of Kiel fjord in February reflected the spring bloom of diatoms in late February and March. Surface sediment biogenic silica content clearly reflects spatial differences in surface water primary productivity, and at low depths and under relatively high sedimentation rates, it could refer to seasonal changes of primary productivity (Rathburn et al., 2001; Bernardez et al., 2006). At the same time, Schwentine river might also be a source of opal for the inner fjord sediments because in the suspension of its water the opal values exceeded 15 wt.% owing to freshwater diatoms. As the maximum of biogenic silica in the inner fjord sediments was not found in the vicinity of Schwentine mouth, we consider the primary productivity in the fjord as the main cause of seasonal and spatial variations in biogenic silica concentrations.

1.4.4 Chlorine and phaeopigments

Chlorine concentrations in surface sediments varied from 7000 to 600 000 mg/g dry sediment (Fig. 1.2). The values were generally higher in March than in December. The spatial distribution of chlorine concentrations was irregular. In December and February, the highest concentrations were observed in the innermost fjord, while in March and May the chlorine levels were elevated towards the outer fjord. This pattern seems to depend on the development of the spring bloom, sequential growth of different algal groups and changes in hydrographical conditions (Graf et al., 1982) as well as terrigenous input. The ratio of chlorophyll *a* to phaeopigments generally increased from February to May, which infers a flux of fresh organic matter to the sea floor (Greiser and Faubel, 1988; Reuss et al., 2005).

1.4.5 Trace metals pollution

The concentrations of copper, zinc, tin and lead in surface sediments of Kiel Fjord show a high variability (Table 1.1). With a sample thickness of one centimetre and presumable sedimentation rate in Kiel Fjord about 1 mm per year (Erlenkeuser et al., 1974, Balzer et al., 1987), one has to keep in mind that the trace metal concentrations present an average over the last 10 years. The concentrations are significantly positively correlated with the C_{org} contents and negatively correlated with the sand content. The correlation suggests that most of the trace metals are bound to organic matter, that they accumulate in muddy sediments, and that they are winnowed from sandy sediments. In fact, elevated metal levels were recorded in the innermost and central fjord (Fig. 1.3). Moreover, exceptionally high metal concentrations were found in surface sediments close to Lindenau shipyards at Friedrichsort, and at Tirpitzhafen Navy base.

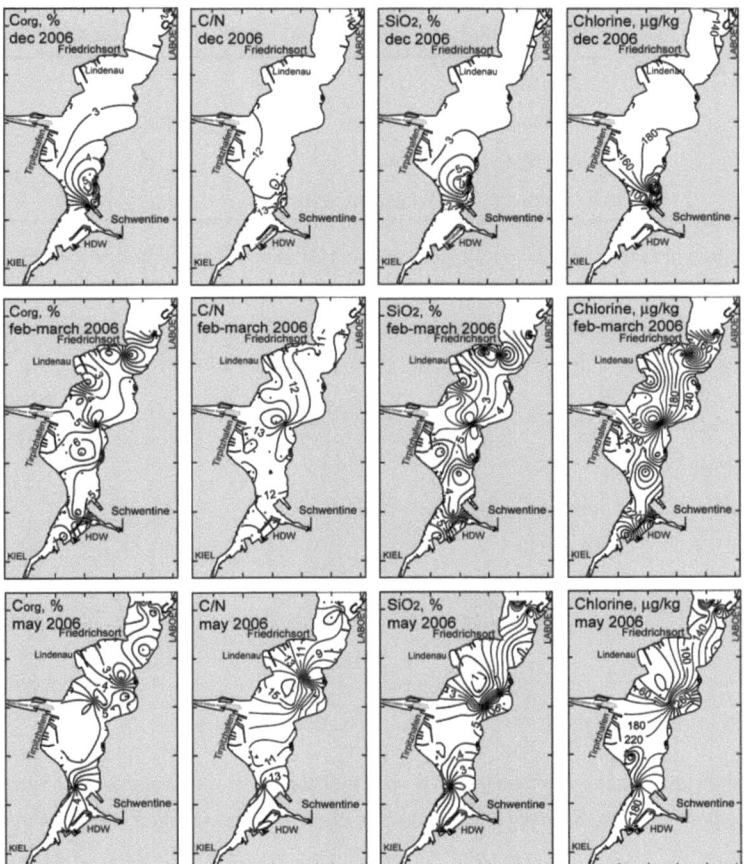

Figure 1.2: Seasonal distribution of organic carbon (%), biogenic silica (wt.%) and chlorine (note: mg/g instead of ng/g by other authors) in Kiel Fjord. Sampling stations are shown here as black dots.

Table 1.1: Mean (range) concentrations of trace metals (Cu, Zn, Sn, Pb, mg/kg) in the surface sediments of Kiel Fjord and their correlation with organic carbon content and sand (>63μm) percentage, number of samples n=53.

Trace metals, mg/kg	Kiel Fjord, mean (range)	Correlation coefficient (r) with C_{org}, %	Correlation coefficient (r) with sand content (>63 μm), %
Cu	62.3 (1.79 - 162)	0.726	-0.581
Zn	185 (11.2 - 434)	0.770	-0.621
Sn	4.97 (0.24 – 18.4)	0.549	-0.404
Pb	118 (6.81 - 260)	0.675 (n=52)	-0.579 (n=52)

The long history of human impact in Kiel Fjord suggests that metal concentrations are substantially higher than the regional background (HELCOM, 1993). Except in the innermost fjord, trace metal concentrations are well in the range of values reported from elsewhere in Kiel Bight for the years 1999 to 2004 (Leipe et al., 1998; Haarich et al., 2003; Pohl et al., 2005).

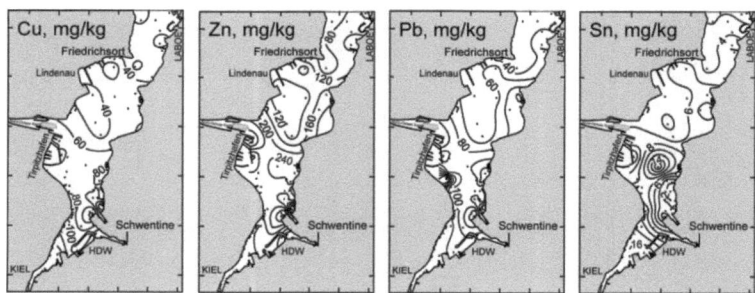

Figure 1.3: Trace metal (Cu, Zn, Pb, Sn) distribution in Kiel Fjord.

Nonetheless, a trace metal study from a sediment core from Kiel Bight demonstrated that the metal concentrations systematically increased since the 1830s and reached maximum in 1950-70s (Erlenkeuser et al., 1974). The youngest Cu, Zn and Pb contents were estimated as 70, 230 and 80 mg/kg respectively. We found the average values of 62, 185 and 118 mg/kg Cu, Zn and Pb. As such, no significant changes in heavy metal concentrations took place during last 40 years.

To the north of Kiel Canal, we found even lower concentrations than in the 1960s, presumably referring to environmental protection measures, in particular, a banning of lead additives in gasoline during the last decades. This may explain the today's low concentrations of lead in Kiel Fjord keeping in mind that its main sources in the Baltic region are atmospheric input and surface runoff (Brügmann, 1996). Tin concentrations were not reported in early investigations. In Kiel Fjord the concentration of tin in the sediment fraction <2000 µm (LANU archive: Ostseemonitoring Programme) was 24 mg/kg in 2004 whereas in other fjords and bays of Kiel Bight it varied from 4 to 17 mg/kg. Our measurements range from 0.2 to 18 mg/kg and confirm the elevated levels in the inner fjord. This can be related to sport harbours and shipyards despite the recent restriction of tin-containing antifouling paints (IMO, 2005).

1.4.6 Foraminiferal population density and species composition

The foraminiferal population density in Kiel Fjord ranged from 3 to 4895 ind/10cm^3, on average 200 to 400 ind/10cm^3. The living benthic foraminiferal communities were dominated by *Ammonia beccarii* (52% on average) and subspecies of *Elphidium excavatum* (together 44% on average). *Elphidium incertum, Elphidium albiumbilicatum* and *Elphidium gerthi* were common (5.3 and 3% on average). *Ammotium cassis, Reophax dentaliniformis regularis, Elphidium williamsoni*, and *Elphidium gunteri* were rare (maximal 2%). The stations with a predominance of *A. beccarii* generally have a lower abundance of *E. excavatum excavatum* and vice versa. We do not recognize any physical, biological or chemical parameter that would explain this spatial change in dominance. But we cannot entirely rule out that these species occupy different ecological niches. As such, we can presume a substitution of these species. *E. incertum* and *E. albiumbilicatum* co-occurred with moderate abundances to both sides of Friedrichsort Sound. *E. gerthi* and *E. williamsoni* were recorded in shallow and near-shore samples (Fig. 1.4).

The arenaceous species *R. dentaliniformis regularis* and *A. cassis* were recorded only sporadically in our samples. The situation was quite different in the 1960s, for instance, Lutze (1965) reported *A. cassis* with up to 2% of the living fauna in Kiel Fjord (Fig. 1.5b). We re-examined 4 of Lutze's samples curated at the Institute of Geosciences (University of Kiel) and revisited his stations in February 2006 (Fig. 1.5a, c). The samples taken in 2006 revealed a 5 to 445-fold increase of foraminiferal population densities as compared to the 1960s. We also did not find *A. cassis*. This species was common elsewhere in Kiel Bight until the mid 1990s (Schönfeld and Numberger, 2007a). Our results infer that *A. cassis* has apparently disappeared in the 2000s from Kiel Fjord too, and that it has been presumably replaced by *A. beccarii*.

Chapter 1 - Foraminiferal response

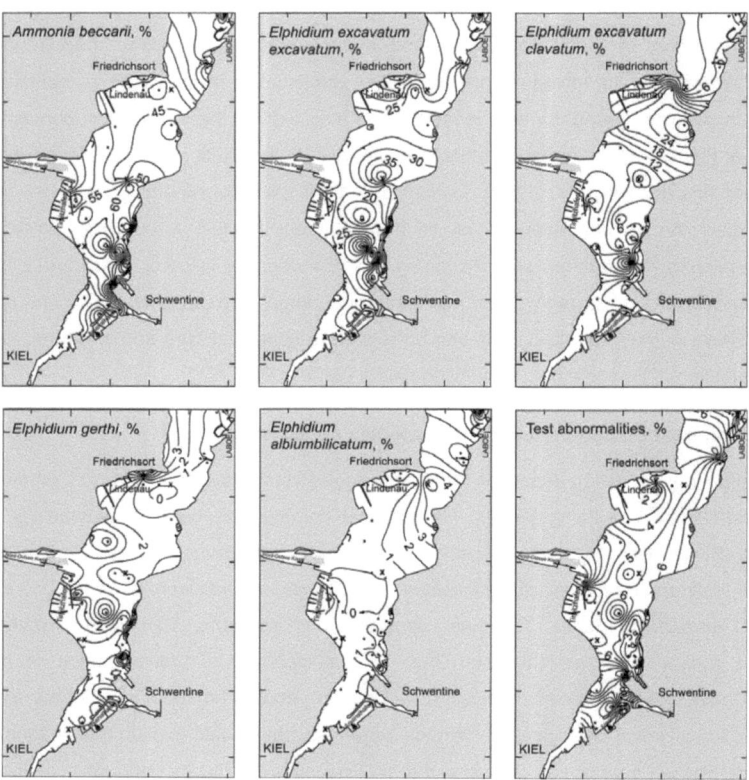

Figure 1.4: Foraminiferal relative abundances and test abnormalities percentage in Kiel Fjord, here X indicates the stations revisited after Lutze (1965).

Positive correlations of population density with biogenic silica (r=0.475; n=21) and chlorophyll a (r=0.600; n=21) were found for samples taken in December. This underpins the strong relationship of the availability of food, in particular diatoms, and foraminiferal population density (Altenbach, 1992; Schönfeld and Numberger, 2007b). A correlation matrix of all geochemical and foraminiferal data is given in Appendix 2.4.

In order to reveal the stress response capability of the benthic foraminiferal fauna, we calculated the ration of the tolerant species *A. beccarii* vs. the specialized *E. excavatum* (A/E Index), firstly described by Sen Gupta et al. (1996) as a proxy of hypoxia. The highest A/E values were found in the central part of Kiel Fjord. They coincide with high C_{org} (7 %) and tin concentrations (18 mg/kg).

In the inner part of Kiel fjord, we recorded high frequencies of test abnormalities (up to 17 %). This is considerably higher than the typical value of 1 % under natural undisturbed conditions (Alve, 1991; Yanko et al., 1999). The majority of abnormal tests were observed in *A. beccarii*. A high number of test abnormalities preferentially occurred in the inner fjord, where the highest trace metal levels were marked.

E. albiumbilicatum has been described as a typical shallow-water species (Lutze, 1965). Here, it inhabits the transitional area of Friedrichsort Sound where sandy sediments prevailed. The high water turbulences seemingly prevent the accumulation of organic matter bound trace metals here. On the other hand, it was suggested that species living in turbulent waters develop spines (Boltovskoy et al., 1991). Tests of *E. albiumbilicatum* possess the numerous pustules in apertural and umbilical areas (Plate 1,Fig. 20) making the test surface rough and enabling this species to withstand the higher water turbulences in this sound.

The species composition of dead assemblages at stations revisited after Lutze was the same as the living assemblages. Lutze (1965) reported that thanatocoenoses in the 1960s also resembled the living communities. In 2006, the living/dead ratios varied from 0.3 in the inner part to 3.2 in the outer Kiel Fjord, which is on average 5 times higher, than it was in the 1960s (Appendix 2.3).

The remarkable increase in population densities as compared to previous studies in Kiel Fjord arises a question: why living foraminifera became so abundant since the 1960s, especially in the presence of trace metals? According to Yanko et al. (1999), some foraminifera might respond positively when the environmental impact is continuous. On the other hand, there are no data on trace metal concentrations in Kiel Fjord from the 1960s and therefore one cannot conclude that trace metals are the only factor that is responsible for the observed changes. Moreover, after the setup of sewage treatment plants and strict environmental protective politics in the 1990s (e.g. Danish Action Plan (I); HELCOM), which caused a decrease of industrial discharges and agricultural load, a general decrease of nutrient inputs and stabilization of oxygen levels in the SW Baltic took place (Nausch et al., 2003a, 2004). Despite the slight decline in nutrient levels since the mid 1990s, an increase of primary production by roughly 40% during the past 30 years has been suggested for the western Baltic Sea (Wassmann, 1990; Schönfeld and Numberger, 2007a).

Provided this is applicable for Kiel Fjord too, even a doubling in primary production can not explain a 67-fold increase in foraminiferal population densities from 23 ind/10cm^3 on average in 1963 to 1582 ind/10cm^3 on average in 2005 and 2006.

1.4.7. Re-examination of Lutze's material

Differences in results shown at Fig. 1.5a and 1.5b may refer to discrepancies in taxonomy, sampling seasons, size fractions (> 63 µm in this study and >100 µm by G.-F. Lutze) and study of the whole samples (in 2006) vs. concentrates (1960s). Fig. 1.5b shows *E. excavatum* subspecies, lumped together in 1960s, as the dominant elements of the living fauna. *E. incertum* had higher abundances, whereas *A. cassis* and *R. dentaliniformis* were rare. Lutze did not report *E. albiumbilicatum* and *E. gerthi*, which we found in his samples. Apparently he recognized both species as variants of *E. excavatum*. Lutze's sampling campaign started in spring 1962 and continued until fall 1963. Regarding the difference in sample numbering (342 vs. 239), it well might be that sampling in the 1960s also comprised several seasons per year, as we did in the current study. Concerning the differences in size fractions, it was shown that there were no living specimens smaller than 80 µm observed in the western Baltic Sea (Schönfeld & Numberger, 2007a: p.85). Therefore it is unlikely that G.-F. Lutze missed or washed away a significant proportion of the fauna. Most residues of Lutze' samples contained a very few or no living specimens whereas the respective flotation concentrates were very rich. Therefore, even if Lutze examined only concentrates but not the whole samples, his results on the population density would not differ by two orders of magnitude to the results we obtained in our 2006 survey. Thus, we finally consider the differences in methods in this study comparing to the 1960s to be of minor influence on the final result.

1.4.8. Invasion and opportunistic behaviour of *Ammonia beccarii*

A. beccarii has an ubiquitous distribution in the Kiel Fjord whereas both *E. excavatum* subspecies show avoidance of the central fjord with silty sediments enriched in C_{org} and tin. In the North Sea, Sharifi (1991) described *E. excavatum* as more frequent than *A. beccarii* in sediments polluted by Zn. According to Alve (1995), abundant and geographically widespread species are to be considered as most tolerant to environmental pollution. *A. beccarii* is commonly frequent in coastal and paralic environments (e.g. Stouff et al., 1999ab). Taking all this into account, we consider that the main reason why *A. beccarii* is so abundant in Kiel Fjord, its opportunistic behaviour and high potential to survive under high input of nutrients and trace metal concentrations.

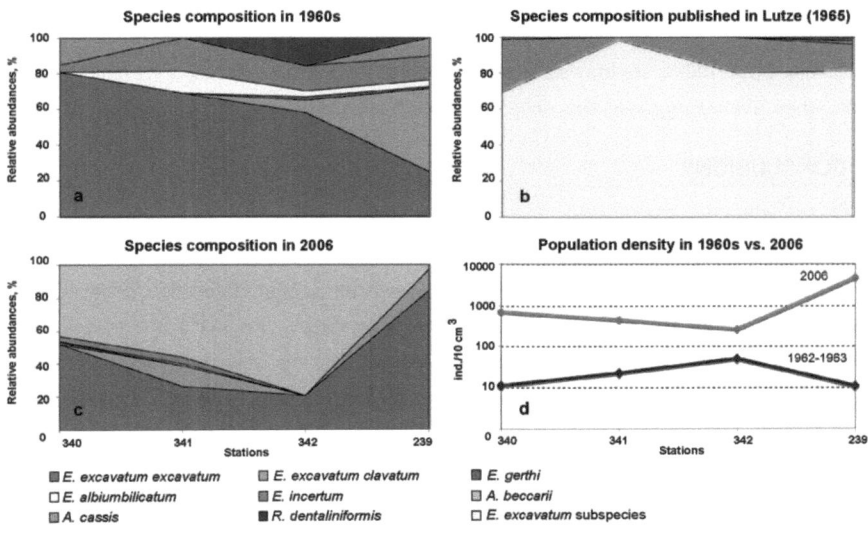

Figure 1.5: The species composition in re-examined samples from Lutze collection taken in 1962-1963 (a), described by Lutze himself (b) and at stations revisited after Lutze in February 2006 (c). Figure (d) shows the changes in population density (log scale) since 1960s (Lutze, 1965).

1.4.9. Disappearance of *Ammotium cassis*

Sample PO220-37-2 taken in the Kiel Bight in 1996 had 90% of *A. cassis*, but did not contain any calcareous foraminifera. It may well be, that due to bad storage conditions, all the calcareous tests were dissolved in this sample. For this reason, we revisited station PO220-37-2 in December 2005 but we did not find living *A. cassis* any more. Lutze (1965) stated that foraminifera in the Baltic Sea are mainly salinity and temperature dependant, and that *A. cassis* is adapted to a strong halocline between surface an deep waters in Kiel Bight. Schönfeld and Numberger (2007a) suggested cyclic changes of *A. cassis* abundances depending on saltwater inbursts in the Kiel Bight and high salinity contrasts between surface and deep waters. As we observed only isolated specimens in some places, the inner Kiel Fjord is currently almost unpopulated by *A. cassis*. This pattern may be due to the fact that the inner fjord is shallower, more closed and less saline than the open Kiel Bight. As such, the deep boundary layer, which is a necessary condition for nutrition of *A. cassis*, cannot establish in Kiel Fjord (Olsson, 1976).

It is conceivable that with faunal change from very large *A. cassis* to much smaller *A. beccarii*, the total biomass might decreased. However, as the population density increased significantly since 1960s, we may assume that biomass today is higher than it was in 1960s and 1990s, when *Ammotium cassis* was abundant in the Kiel Bight.

1.5 CONCLUSIONS

The results of the present study showed that labile organic compounds (biogenic opal, chlorines, C_{org}) in sediments of the Kiel Fjord were subjected to a strong seasonal variability. Their concentrations are significantly higher in springtime. The spatial distribution of labile organic compounds is mainly determined by sediment type. Generally, the levels of concentrations of biogenic compounds are comparable to those reported from the open Kiel Bight. Markedly low levels of food in Friedrichsort Sound establish quite unfavourable conditions for many benthic foraminiferal species. The surface sediment pollution by copper, zinc, tin and lead principally could be considered as moderate because the levels of metals are comparable to those elsewhere in the Baltic Sea. Nevertheless the inner Kiel fjord is distinguished by a high load of heavy metals. The high tin concentrations in surface sediments apparently depend on its accumulation in muddy sediments for previous decades.

The analysis of foraminiferal population density shows a patchy distribution and a response to food availability, which is depicted by SiO_2 and Chl *a* in the sediments. The strong increase of population density since the 1960s remains enigmatic. It cannot be attributed to an increase in organic matter supply and a slight reduction of pollution. Furthermore, we observed significant changes in foraminiferal species composition in 2005-2006 as compared to the 1960s. The stress-tolerant species *A. beccarii* invaded Kiel Fjord. We suppose that this species is highly opportunistic and can tolerate elevated levels of nutrients and trace metals. *E. albiumbilicatum* apparently is able to withstand the higher water turbulences and therefore inhabits the transitional area of Friedrichsort Sound. Unfavourable salinity conditions in the Kiel Bight and absence of a deep halocline in Kiel Fjord might have caused the disappearance of *A. cassis* during the past decades.

ACKNOWLEDGEMENTS

The authors thank Wolfgang Kuhnt who provided access to original G.-F. Lutze's samples at the Institute of Geosciences at Kiel University and samples from RV Poseidon cruise 220. We are also grateful to Joachim Voß and Thorkild Petenati (LANU), who made the unpublished data of LANU archives available. We kindly appreciate the technical assistance of Jutta Heinze, Anna Kolevica,

Bettina Domeyer, Anke Bleyer, Udo Laurer (IFM-Geomar), Ute Schuldt, Claudia Ehlert and Ulrike Westernströer (Institute for Geosciences, Kiel). We acknowledge the crew of RV Polarfuchs Holger Meyer and Helmut Schramm for help with sampling. Our sincere thanks goes to Elisabeth Alve (University of Oslo) for her valuable advice on early stage of this study. We are grateful to Frans Jorissen and Peter Frenzel for constructive review comments. This study was funded by German Academical Exchange Service (DAAD) and Leibniz Award DFG DU 129-33 of Prof. W.-Ch. Dullo.

Chapter 2

Recent benthic foraminifera in Flensburg Fjord: distribution and response to environmental change

Published as:
Polovodova, I., Nikulina, A., Schönfeld, J., and Dullo, W.-Ch., 2009. Recent benthic foraminifera in the Flensburg Fjord (Western Baltic Sea), *Journal of Micropaleontology*, 28: 131-142.

ABSTRACT

Living benthic foraminifera of Flensburg Fjord were surveyed in June 2006. The muddy and organic-rich sediments of the inner fjord were dominated by *Elphidium incertum*. Species *E. incertum,* and *E. excavatum* were frequent in muds and sandy muds of the fjord loop around Holnis Peninsula and in the outer part. Gelting Bay yielded a different biofacies indicating a brackish and sandy habitat; poor in food supply, and with microfauna dominated by *Ammonia beccarii* and *E. albiumbilicatum*. The central fjord and near-shore zones of the loop were characterised by sandy muds, relatively poor in food and occupied by *A. beccarii, E. incertum* and *E. excavatum* subspecies. High abundances of *E. excavatum* were encountered in the innermost fjord with fine-grained and organic-rich muddy sediments.

A comparison with previous studies revealed the profound changes in species composition in the outer Flensburg Fjord since the 1970s. A decline in numbers of *Ammotium cassis* and flourishing of *Ammonia beccarii* in Gelting Bay were recognised. These changes were most likely associated with decreased intensity and frequency of salt-water inflows into the Baltic Sea since the 1960s. It is inferred that *A. cassis* decline is similar to that of *Eggereloides scaber*, which currently is found only in depressions of Kiel Bight with higher salinity.

2.1 INTRODUCTION

Foraminifera in the Baltic Sea have been investigated since the 19th century (Schulze, 1875; Möbius, 1889; Levander, 1894). First studies with a taxonomical identification of foraminifera in Kiel

Bight were accomplished by Rhumbler (1936), who found 49 genera and 71 species. Studies on foraminiferal ecology in the SW Baltic Sea, initiated by Rottgardt (1952), established temperature and salinity as the most important controlling ecological factors for benthic foraminifera, whereas substrate was assumed to be of minor significance (Lutze, 1965). Later, previously ignored substrate features together with hydrodynamics and oxygen content of the bottom water were brought into consideration by Wefer (1976), who found them to be important factors regulating foraminiferal abundances in the open Kiel Bight. Some species in the central and south-western Baltic were shown to be especially sensitive to salinity changes caused by saline bottom water inflows from the Kattegat (Lutze et al., 1983; Hermelin, 1987; Schönfeld & Numberger, 2007a; Nikulina et al., 2008). In particular, Lutze et al. (1983) observed an arenaceous foraminifer *Eggerelloides scaber* only in depressions of Kiel Bight due to its preferences for higher salinity. For the central Baltic Sea, it was shown that a northward decrease in salinity, temperature and oxygen content is reflected by reduced benthic foraminiferal diversity and abundances (Hermelin, 1987). Kreisel & Leipe (1989) described only four species in the Bay of Greifswald and suggested that such low species richness also might be explained by brackish conditions. A decrease of salinity as a background reason for the decline of *Ammotium cassis* across the Kiel Bight was suggested by Schönfeld & Numberger (2007a) and Nikulina et al. (2008). These authors assumed that reduced salinity prevents formation of a stable halocline, which is necessary to provide a high supply of suspended food particles that are essential for the feeding of *A. cassis* (Olsson, 1976).

Lack of food is generally considered as one of the main constraints on foraminiferal abundances (Murray, 2006). In Kiel Bight, limited food availability was observed to be of major importance for only two species: *Ammotium cassis* and *Ophthalmina kilianensis*, whereas all other foraminifera had enough food due to the high productivity in the SW Baltic as compared to the adjacent ocean (Wefer, 1976). Feeding behaviour of foraminifera in Kiel Bight was studied by Heeger (1990) and Linke & Lutze (1993), who also reported some adaptive mechanisms to gain more food under less favourable conditions, like elevated microhabitats, change from epifaunal to infaunal mode and isolated chloroplasts in the protoplasm assumed as an additional source of energy. Schönfeld & Numberger (2007b) reported the "bloom-feeding" strategy of *E. excavatum clavatum* in Kiel Bight, reflected in elevated pigment content in the protoplasm.

In spite of numerous investigations in the SW Baltic, foraminifera from the entire length of Flensburg Fjord (Fig. 1) have not so far been studied and this is the first description of living (stained) benthic foraminiferal assemblages from this area. Previous investigations either considered the total assemblages or were focussed only on the central, or open part of the fjord

(Rottgardt, 1952; Exon, 1972). The inner fjord has not previously been considered. Because Flensburg Fjord is an inlet of Kiel Bight, which directly faces the salt-water inflows from the Kattegat, some foraminiferal taxa dwelling here may be very sensitive to occasional salinity fluctuations (Polovodova & Schönfeld, 2008), while others may be well adapted. Therefore, Flensburg Fjord represents an appropriate study area for dynamic response of foraminifera to natural environmental perturbations.

The purpose of this study is to investigate the distribution of living (stained) benthic foraminifera in Flensburg Fjord with respect to environmental parameters of the bottom water and sediments. Further, these data are compared with previous studies in order to assess the changes in foraminiferal communities since the late 1940s and 1970s.

2.2 STUDY AREA

Flensburg Fjord is a narrow, 50 km-long, W-E trending inlet of the north-western part of Kiel Bight (Fig. 2.1). The inner fjord, which is 10-20 m deep is characterised by restricted water exchange with the Kiel Bight and the Baltic Sea over a sill of 10 m depth off Holnis Peninsula (Nikulina & Dullo, 2009). The outer Flensburg Fjord comprises Sonderborg Bay with 13-31 m depth range, Gelting Bay (4-22 m deep) and open waters to the east of Gelting Peninsula. There is a considerably high depth range – from 5 m at Schleisand up to 39 m in the Little Belt.

Flensburg Fjord is the most protected estuary in the region and wave action does not reach significant depth. Thereby, waves and currents play a major role in the shallow-water areas. They favour the erosion of cliffs and deposit the eroded material in the deeper waters of the outer fjord (GKFF, 1973). Destruction and transport of foraminiferal tests are of minor importance in sheltered depositional areas as compared to regions with enhanced sediment erosion. The sediments of the deeper basins in the outer fjord are dominated by sandy muds and silt, whereas in shallow coastal areas sand and muddy sand prevail (Exon, 1972; GKFF, 1973). The inner Flensburg Fjord has sediments composed of dark sandy mud and soft mud (Exon, 1972).

During winter months, the inner fjord is well mixed at 6.5 °C and shows a salinity of 21 psu. The water column has a pronounced stratification during summer. Surface water of 17.5 °C on average and a salinity of 16.5 psu overlies the bottom water of 11 °C and 18 psu. Every summer, a stable thermocline develops at 8-9 m in the inner fjord (GKFF, 1973). In the outer fjord, a persistent pycnocline at 16-20 m (Exon, 1972) separates brackish surface water from salty deep water throughout the year. The upper boundary of the pycnocline coincides in the outer fjord with the

Chapter 2 – Recent benthic foraminifera

effective depth of wave action and divides the sedimentary environments into non-depositional and depositional areas.

The stable stratification in the inner part, together with enhanced eutrophication in the 1970-80s, was responsible for summer oxygen deficiency, which lasted several months without interruption (Wahl, 1985). Eutrophication of Flensburg Fjord was caused by high nutrient input from sewage outfalls (Rheinheimer, 1970) and fertilizers from the adjacent land (DDTFF, 1992; LANU, 2001). In spite of the absence of nutrient input since that time, high levels of nitrogen and total organic carbon have remained in the sediments (LANU, 2001).

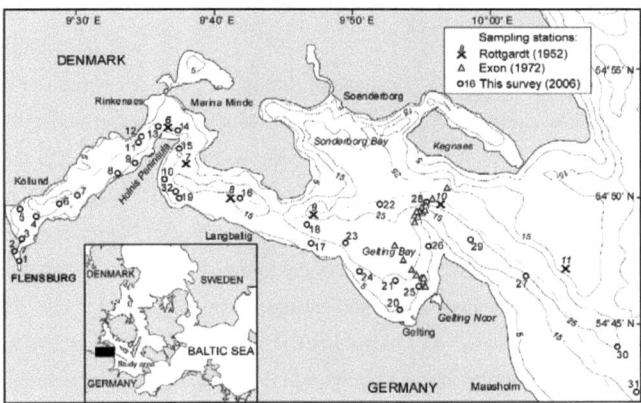

Figure 2.1: Regional setting of Flensburg Fjord and positions of sampling sites. Black box indicates the location of study area within the SW Baltic. Prefixes "PF16-" at every station are not given for sake of convenience. For longitude, latitude and names of Exon (1972) samples see Appendix 2.

2.3 MATERIAL AND METHODS

Thirty-two samples of the surface sediment were taken on 2 daily cruises of R/B *Polarfuchs* in June 2006. An access was limited to German territorial waters and therefore concentrated on the southern part of the fjord. The majority of the samples were retrieved with a Rumohr corer, which has a sampling tube of 56 mm inner diameter. A Van-Veen Grab of 250 cm^2 surface area was used when sandy sediments were encountered.

Within minutes of sample retrieval, salinity, temperature and dissolved oxygen content of bottom water were measured on board with Oxi- and Conductivity meters (WTW Oxi323/325 and LF320). In order to avoid the bias produced by spatial patchiness in the range of 2 m^2 in foraminiferal distribution (Lutze, 1968), the Rumohr corer was always deployed three times and the uppermost 1 cm of sediment was taken off on each deployment (Schönfeld & Numberger, 2007a, b). This sampling strategy was based on the previous observations in the SW Baltic Sea (Wefer, 1976) that more than 90% of living foraminifera are concentrated in first centimetre of sediments. When sampling with a Van-Veen Grab, cut-off syringes marked with a centimetre-scale were gently pushed vertically into the sediment to sample the top centimetre. The samples from all three deployments (about 25 cm^3 of wet volume each) were collected in a glass vial, carefully mixed and subsampled for organic and inorganic geochemical analyses. The remaining 45 cm^3 of sediment were transferred to a PVC vial, then preserved and stained with a Rose Bengal – ethanol solution (2g/l) for foraminiferal studies.

The foraminiferal samples were first passed through a 2000 μm screen in order to remove bivalve shells or pebbles, and, then they were gently washed with tap water through a 63-μm sieve. Both fractions (63-2000 μm and >2000 μm) were dried at 60°C. Only well-stained foraminifers, presumed as living at the time of sampling (Murray & Bowser, 2000), were picked out from respective aliquots, sorted at the species level and mounted in Plummer cell slides with glue. Dead assemblages were not considered in this study to avoid taphonomic bias. The main species were photographed using scanning electronic microscopes JSM-6460LV (St. Petersburg State Mining Institute) and Cam Scan CS-44 (Institute of Geosciences, Kiel University). In comparison to Cam Scan CS-44, JSM-6460LV allows to observe the natural test surface without electric „noise" at images due to operation at low vacuum (30 Pa) with no metal or graphite coating of the samples (Widerlund & Andersson, 2006). Graphite glue was used to mount the foraminiferal tests on the aluminium stabs prior to observation with JSM-6460LV, whereas carbon adhesive tape was applied for sample preparation for the Cam Scan.

To avoid staining errors, at least 100 specimens were picked either dry or wet and counted as a representative sample for ecological studies with satisfactory reliability (Murray, 2006). Wet picking was applied in sand-rich samples to facilitate the recognition of stained specimens.

For the species proportions, the standard binomial error was calculated (Fatela & Taborda, 2002). In order to avoid bias associated with a constant sum constraint due to the low number of species in our samples, the percentage data were log-ratio transformed, following Kucera & Malmgrem (1998). After data transformation, Q-mode cluster analysis was performed to distinguish different groups and biofacies of living benthic foraminifera. The statistics software Statistica 6.0 was used, and only the most abundant foraminiferal species with abundances higher than 5% were considered in the cluster analysis. The applied clustering method was complete linkage and

Euclidean distances (Vance et al., 2006). The resultant groups of samples are considered to represent different biofacies. The groups were defined by taking into account the significance level of linkage distance. We also considered the average percentages of species within each sample group in order to discern the different ecological groups of foraminifera, which characterise a certain biofacies.

The geochemical data are reported in details by Nikulina & Dullo (2009) and are used here for comparison only.

2.4 RESULTS

2.4.1 General trends in foraminiferal distributions

Living populations and surface-sediment assemblages are useful tools to assess the current state of a benthic ecosystem (Hallock, 2000). In case with benthic foraminifera, ecological studies should be based on living (stained) assemblages (Murray, 2006), because dead or total assemblages do not reflect the current environmental conditions due to taphonomic alterations. In order to describe the current foraminiferal distribution and ecology in Flensburg Fjord and to assess the response of these organisms to the recent environmental changes, only living (stained) foraminifera were considered in this study. Ten taxa (8 calcareous, 2 arenaceous) were identified at 31 of 32 stations; one station (PF15-09) was barren of living foraminifera.

The standing stock of benthic foraminifera varied in a high range: from 11 to 3130 individuals per 10 cm^3 (Fig. 2a). Numbers lower than 20 living specimens per 10 cm^3 occurred in the innermost fjord and off Gelting Peninsula. The highest standing stock values of more then 3000 individuals per 10 cm^3 were recorded in the outer part of the Flensburg Fjord.

Assemblages were dominated by *Elphidium incertum, Ammonia beccarii,* and *E. excavatum excavatum* (28, 25 and 25 % on average). *E. excavatum clavatum* and *E. albiumbilicatum* were common (12 and 10 % on average). *Ammotium cassis, Reophax dentaliniformis regularis, E. williamsoni, E. gerthi,* and *E. gunteri* were rare (maximal 2%). The inner fjord was characterized by a predominance of *E. incertum* (Figs 2.2c, e) with a small proportion of *E. excavatum excavatum*. The foraminifer *E. albiumbilicatum* was encountered in high abundances (up to 56 %) in near-shore sandy areas of Gelting Bay (Fig. 2.2f) with enhanced coastal erosion and active sediment transport. *Ammonia beccarii* occurred in the central fjord and Gelting Bay (Fig. 2.2b) under conditions of reduced food availability. Both subspecies of *E. excavatum* showed the highest abundances in the open Flensburg Fjord (Figs 2.2c, d). Arenaceous species *Ammotium cassis* and

Chapter 2 – Recent benthic foraminifera

Reophax dentaliniformis regularis tended to occur at stations with higher salinity, situated in the central and open parts.

Significant correlations between certain species, sediment type and food supply were revealed (Table 2.1). In particular, *Ammonia beccarii* and *Elphidium albiumbilicatum* correlated positively to the proportion of the sediment fraction >63µm (sand content). Conversely, abundances of *E. excavatum excavatum* showed a negative correlation with sand content. In its turn, *A. beccarii* showed a negative correlation to food availability.

Table 2.1: Correlations between species abundances and environmental parameters in Flensburg fjord. The significance-test for a linear correlation at normal distribution of data was performed according to Aßmann and others (2007). Note that a type I error $\alpha=0.05$. All correlations, presented here, passed significance-test i.e., have T-value in modulus higher than $t_{n-2, 1-\alpha}$.

Species	Environmental parameter	Correlation coefficient, (r)	Significance test, (T)	Quantile, $(t_{n-2, 1-\alpha})$
Ammonia beccarii	Fraction >63 µm, (%)	0.484 (n=31)	2.979	1.699
-	SiO_2, (weight %)	-0.684 (n=31)	-5.049	1.699
-	TN, (%)	-0.708 (n=30)	-5.305	1.701
-	Chlorophyll a, (ng/g)	-0.525 (n=31)	-3.332	1.699
	Phaeopigments, (ng/g)	-0.700 (n=31)	-5.279	1.699
E. excavatum excavatum	Fraction >63 µm, (%)	-0.405 (n=31)	-2.385	1.699
E. albiumbilicatum	Fraction >63 µm, (%)	0.417 (n=31)	2.471	1.699

SEM images of foraminifera showed that some tests of *Elphidium incertum* were covered with particles, similar to sand grains (Plate 2, Fig. 14-15). *Elphidium albiumbilicatum* and *E. excavatum clavatum* encountered in Gelting Bay showed irregular mineralogical projections of unknown origin, which protruded from the test wall (Plate 2; Fig. 7, 19). An intense perforation of *Ammonia beccarii* tests (Plate 2; Fig. 22) were found in central (PF16-21) and southern Gelting Bay (PF16-20), and in the open Flensburg Fjord (PF16-27). *Elphidium albiumbilicatum* showed numerous pustules in the apertural and umbilical area (Plate 2; Figs 17-19), which enable it to tolerate the intensive hydrodynamics due to a rough test surface (Nikulina *et al.*, 2007).

Chapter 2 – Recent benthic foraminifera

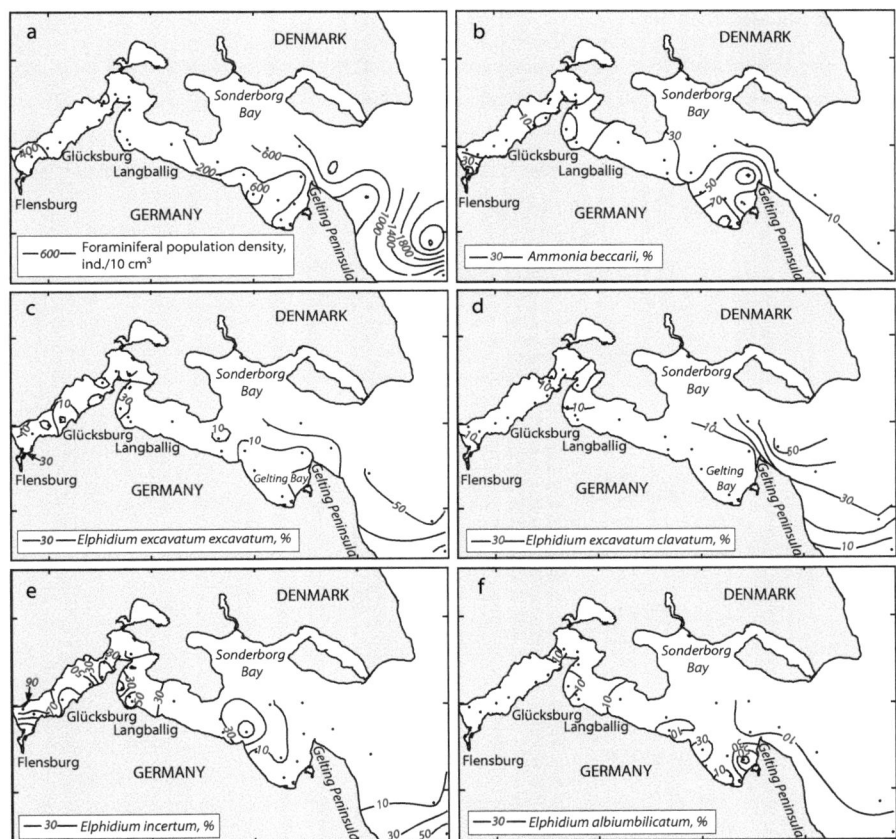

Figure 2.2: Benthic foraminiferal standing stock (a) and relative abundance of the dominant species (b-f) in the Flensburg Fjord. Black dots indicate sampling locations.

2.4.2 Cluster analysis

The dendrogram (Fig. 2.3) is composed of several nesting groups, which are here grouped into 5 biofacies (Table 2.2 & Fig. 2.4). Biofacies 1 (*Elphidium incertum*) is associated with muddy sediments and high food concentrations. Despite this biofacies is close to the limit of significance level, we considered it as one cluster based on relative abundances of the dominant species, which were higher than 50 % at all stations.

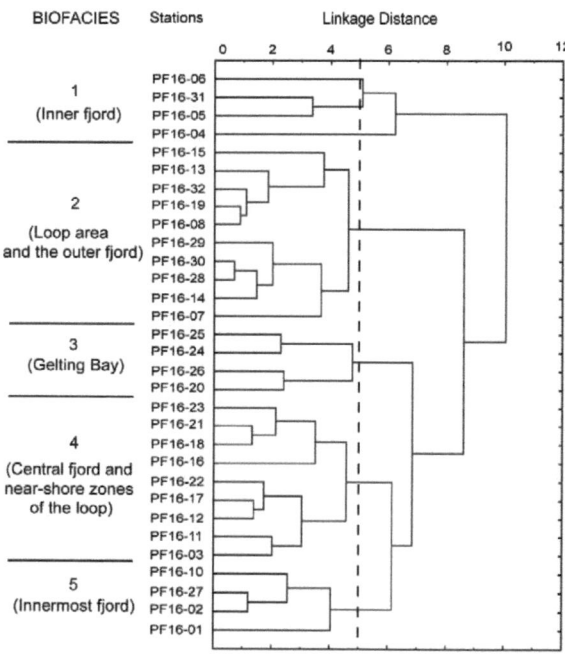

Figure 2.3: Dendrogram produced by cluster analysis (complete linkage, Euclidean distance) of log-transformed percentages of living foraminifera. Dashed line indicates the significance level of the Euclidean distance, according to a scree-plot.

Biofacies 2 (co-dominant *E. incertum* and *E. excavatum excavatum*) occurred in muddy sediments with high chlorophyll a content. Biofacies 3 (*Ammonia beccarii*) occurred in coarse sands poor in food supply and overlain by highly oxygenated water. Biofacies 4 (*A. beccarii* with *E. incertum* and *E. excavatum excavatum*) occurred in muddy sands. Biofacies 3 and 4 had differences in the proportions of the subsidiary species. Biofacies 5 (*E. excavatum excavatum*) occurred in muddy sediments. There were minor variations in species richness between the biofacies but major variations in both the average standing stock and the range of standing stock. This was especially recognised in biofacies 3 where it ranged from 28-200 ind. 10 cm^{-3} in the loop area and 938-3130 ind. 10 cm^{-3} in the outer fjord. Water temperature and salinity differences showed only minor variation between the biofacies.

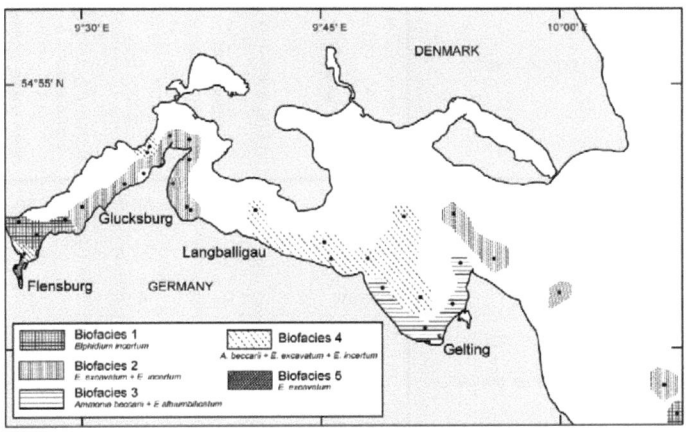

Figure 2.4: Biofacies distribution of living benthic foraminifera in Flensburg Fjord, as distinguished by sample groups of the cluster analysis.

2.5 DISCUSSION

2.5.1 Food supply

As a measure of food supply, we considered the sedimentary content of chlorophyll a, biogenic silica, and organic carbon. In general, the flux of organic carbon reaching the seafloor comprises labile material from recently died plankton, refractory material both derived from plankton and having terrestrial origin, and faecal pellets from zooplankton (Murray, 2006). Nikulina & Dullo (2009) reported the high levels of organic carbon, chlorophyll a and biogenic silica in sediments from the inner Flensburg Fjord, which mirrored phytoplankton bloom deposition. The phytoplankton in this area mainly consisted of diatoms (Schönfeld & Numberger, 2007b). These observations allow us to consider the above-mentioned parameters as a measure of food available for benthic foraminifera in the study area.

Chapter 2 – Recent benthic foraminifera

Table 2.2: Average foraminiferal census data (%) and environmental parameters in each biofacies. Note that only species with abundances higher than 5% are given.

	Benthic foraminiferal species and environmental parameters	Biofacies 1 Inner fjord	Biofacies 2 Loop area and outer fjord	Biofacies 3 Gelting Bay	Biofacies 4 Central fjord & near-shore zones of the loop	Biofacies 5 Innermost fjord
Census data	*Ammonia beccarii*	3.1	7.7	65.6	40.2	12.8
	Elphidium albiumbilicatum	9.9	2.6	26.9	11.3	10.3
	Elphidium excavatum excavatum	8.8	31.9	5.2	19.4	57.9
	Elphidium excavatum clavatum	0.3	21.0		7.9	18.3
	Elphidium incertum	75.7	36.0		20.3	
	Species richness (S)	4	6	4	6	5
	Counted specimens (N).	181	175	118	152	84
	Standing stock, ind. $10cm^{-3}$	197	614	182	264	57
	average (range)	(105-324)	(28-3010)	(11-254)	(83-759)	(21-142)
Bottom water	Temperature, °C	8.3	8.3	12.7	8.6	9.6
	Salinity, ‰	22.3	23.5	18.6	22.5	21.3
	O_2 saturation, %	55.5	58.9	99.0	61.2	58.3
Sediment	Fraction > 63 µm, %	21.8	9.7	87.3	30.7	21.1
	Biogenic SiO_2, %	5.3	4.0	0.8	2.5	4.9
	C_{org}, %	6.7	4.5	0.4	3.5	6.0
	Chlorophyll *a*, mkg/g	84.2	157.1	16.6	96.1	208.2

Established correlations between some foraminiferal species, food supply and sediment type also point out at the preservation capability of sediments. In particular, *Ammonia beccarii* and *Elphidium albiumbilicatum* tended to sands of Gelting Bay, where due to the absence of nutrient recycling from bottom sediments (Exon, 1972) and high hydrodynamic activity, the lower productivity and a lack of food supply take place. Conversely, *Elphidium incertum* was found in the inner fjord where the higher productivity, regular oxygen depletion, restricted current regime and muddy sediments enriched in organic matter were encountered (Nikulina & Dullo, 2009).

These observations strengthen the idea that at least some benthic foraminiferal species of the Flensburg Fjord are food dependant. On the other hand, we didn't found any correlations of standing stock with distribution of organic carbon, chlorophyll *a* and biogenic silica. This relationship is often complicated by the covariance of organic carbon with other factors such as sediment grain size and geochemistry in marginal environments (Murray, 2006). The latter two factors are closely coupled, because sediment grain size determines food preservation and accumulation in the sediments, as well as nutrient recycling and formation of certain redox conditions, which affect productivity in turn.

2.5.2 Reliability of foraminiferal abundances

An accuracy of foraminiferal abundances in sediment samples is usually dependant on sampling protocols and processing of the samples. To get reliable data, more than one sample from the same site has to be taken to avoid the bias associated with patchiness of foraminiferal distribution (Murray, 2006). In this study, the pooling of three replicate core top samples was done in order to minimize the bias induced by spatial patchiness. The scale of patchiness in the study area is in the order of 2 m^2 (Lutze, 1968). With ship motion of 1 to 3 m during station work, a replicate sample was most likely retrieved from a place at the seabed where the foraminiferal population structure differed from the place where the first sample came from. Therefore, we had to combine the replicates in order to obtain samples that were representative for the sampling site on a local scale. As the volumes of the replicates were equal, pooling was considered to create a much more representative sample for foraminiferal and geochemical studies than a later combination of the analyses from the individual replicates. Furthermore, such replicate analyses might have been created from splits of a different size leading to a different contribution of the individual samples.

Staining of foraminiferal protoplasm with Rose Bengal was discussed in literature as a method, which has to be handled with caution because it doesn't allow a distinguishing between living and decaying individuals leading to overestimations of species' abundances (Bernhard, 1988; Corliss &

Chapter 2 – Recent benthic foraminifera

Emerson, 1990). The alternative methods for Rose Bengal are fluorescent probes MTT and Cell Tracker Green (Bernard et al., 2006; De Nooijer et al., 2006), which are more accurate in abundance assessment but usually hamper comparison with previous studies and ignore the foraminifera, which did not survive the sampling (De Nooijer et al., 2008). Cell Tracker Green also requires the immediate processing of a sample with high instrumental (UV-light) and preparational (wet picking under dark conditions) efforts (J. Bernhard, personal communication with JS). Conversely, Rose Bengal staining has been and still is widely used in an increasing number of studies. The method is easy in handling and able to give results which are to 96% correct (Lutze & Altenbach, 1991), if it is applied with a long-standing practice and understanding of limitations (Murray & Bowser, 2000). Therefore, we consider an accurate handling of Rose Bengal staining as being able to produce the reliable data of living (stained) foraminiferal abundances.

2.5.3 Predominance of Elphidium incertum in the inner Flensburg Fjord

In this study, we observed *Elphidium incertum* as highly abundant in muddy sediments of the inner Flensburg Fjord associated with higher concentrations of C_{org} (6-11 %) and biogenic silica (3-10 %) and relatively low sedimentary chlorophyll a (43-189 µkg/g). Some authors (Altenbach, 1985; Gustafsson & Nordberg, 1999) found increased amounts of this species after the spring phytoplankton bloom, which delivered fresh food to the sediment surface. We did not find any correlations of *E. incertum* with biogenic silica, chlorophyll a and C_{org}. Apparently, there have to be other factors, which determined the distribution of this species in Flensburg Fjord. As it was shown for Koljö Fjord, a freshly deposited phytoplankton is of less importance for *E. incertum*, when oxygen conditions become unfavourable (Gustafsson & Nordberg, 1999).

Elphidium incertum was described as an infaunal species, which dwells not exclusively at the sediment surface but down to 3-6 cm (Linke & Lutze, 1993). However, if sediments become uninhabitable due to a very shallow redox boundary, this species may be found at the sediment surface (Wefer, 1976). Rottgardt (1952) mentioned the absence of the uppermost brownish oxidized layer in the sediments at station off the Holnis Peninsula, though bottom waters were sufficiently oxidized with 278 µmol/L in May 1949. During sampling in 2006, oxygen concentration at station PF16-14 was lower (178 µmol/L), though saturation exceeded 50 %. The establishment of a redox boundary occurs in Flensburg Fjord every year due to oxygen deficiency of bottom waters from May to October under the presence of a stable pycnocline at 6-12 m depth (Jarke, 1961; Kremling et al., 1979; Wahl, 1984, 1985; LANU, 2007). Thus, we conclude that high abundances of *E. incertum* within the uppermost sediment layer in the inner Flensburg Fjord were most likely related to oxygen deficiency in sediments just before the sampling time.

Moreover, some authors (Wefer, 1976; Linke & Lutze, 1993) reported encystment of *Elphidium incertum* as a strategy of dormancy during anoxic periods. Gustafsson & Nordberg (1999) observed cocooned *E. incertum* tests at the stations, where *Beggiatoa bacterial* mats indicate the presence of a redox boundary at the sediment surface. Such cysts or cocoons are quite resistant and were sometimes still intact after sample processing (Wefer, 1976). In this study, we observed the remains of the cysts firmly attached to tests of *Elphidium incertum* (Plate 1, Figs 14-15), even though all samples were thoroughly washed. The same phenomenon was mentioned by Exon in his notes on the 1970s foraminiferal survey as a peculiar feature of *E. incertum* tests in the outer Flensburg Fjord. Hence we infer that our sampling campaign had followed a short anoxic period, which led to an extremely high abundance of *E. incertum* in the uppermost sediment layer.

2.5.4 Evidence for oxygen deficiency in sediments of the Gelting Bay?

Though hydrographical data did not indicate an oxygen deficiency in the bottom water during the sampling period, the southern part of Gelting bay was previously reported as a quiet area exposed to relatively low oxygen and nutrient content (Exon, 1972). Moodley & Hess (1992) showed that *Ammonia beccarii* developed an adaptation reflected in higher test porosity under the lack of oxygen. Indeed, we observed larger pores in *A. beccarii* in southern Gelting Bay than anywhere else in the fjord. This may be seen as an adaptation to provide an adequate gas exchange under low oxygen conditions, which might have taken place before sampling. In order to test this conclusion, we calculated the *"Ammonia beccarii – Elphidium excavatum* subspecies" index (A/E Index), used as a proxy of hypoxia for the Louisiana Bight by Sen Gupta *et al.* (1996). It showed the highest values at four stations in the Gelting Bay: PF16-20 (98 %), PF16-21 (91 %), PF16-24 (96 %) and PF16-26 (100 %). According to Brunner *et al.* (2006), an index value > 80 % indicates hypoxic conditions, whereas an A/E < 50 % is typical for clearly oxic samples. Though, the A/E values > 90 % at stations PF16-20 and PF16-21 coincide with encountered enhanced porosity, the sandy sediments and low concentrations of C_{org}, chlorophyll *a* and biogenic silica (Nikulina & Dullo, 2009: Fig. 3) do not provide the pre-conditions for oxygen deficiency in sediments. At the same time, A/E Index is apparently not applicable as an oxygen proxy in an area and a climatic zone, so different from the Mississippi and Louisiana Bight (Brunner, written communication, 2008). In this case, the reason for enhanced porosity of *Ammonia* tests observed only in southern Gelting Bay remains unclear.

2.5.5 Comparison with previous studies.

Abundances of foraminiferal assemblages at six stations sampled in Flensburg Fjord in October 1949 (Rottgardt, 1952) were compared to our abundances at stations located nearby. The comparison was hampered by differences in methodology. Rottgardt (1952) considered the total (living plus dead) foraminiferal fauna from grab samples, comprising several centimetres of sediment, whereas we investigated the living (stained) assemblages in the uppermost sediment layer (0-1 cm). Nevertheless, we can derive information about the occurrence and dominance of different species in 1949 and reveal how the foraminiferal assemblages in Flensburg Fjord changed since that time. Regarding previous studies conducted in Kiel Bight (Wefer, 1976), which proved a minor inter-annual and yearly variability in the proportion of dominant species, it appears to be justified to draw some limited conclusions even in spite of different sampling seasons.

According to Rottgardt's (1952) results, *Elphidium excavatum* subspecies were the dominant element of the fauna in the Holnis Shoal (his station 6), whereas *Ammonia beccarii* and *E. incertum* were common. These results agree with our observations. In the central part of Flensburg Fjord (stations 7, 8, 9) *Elphidium incertum* was a dominant faunal element in 1949, while *Elphidium excavatum* subspecies and *Ammonia beccarii* prevailed in the central fjord in 2006. Samples from the outer fjord (10 and 11) were incredibly rich in tests (>1000) of *Eggerelloides scaber* in 1949. In 2006, however, we did not observe any *E. scaber* at the stations PF16-28 and PF16-27 from the outer fjord. *Elphidium excavatum* subspecies were the dominant elements of the recent foraminiferal fauna here. Also, we found a few living individuals of *Ammotium cassis*, a species that was not reported by Rottgardt in 1949.

In October 1970 benthic foraminifera in the outer Flensburg Fjord were investigated by Exon (1972). He established two profiles of stations in an outer fjord (Fig. 1). The first profile extended from the shallow to the deeper parts of the southern Gelting Bay. The second one was situated in the outer part, eastward of Kalkgrund (a shallow water area around Gelting Peninsula). Exon's (1972) methods were very similar to ours. He worked with grab samples, from which the uppermost centimetre of sediment was removed, stained with Rose Bengal and studied for living (stained) and dead foraminifera. An exception represents the way in which Exon expressed the values of population density: he weighed wet samples and calculated foraminiferal population densities in ind./10 g of wet sediments. This protocol makes our data on standing stock incomparable with those from the 1970s survey. In order to figure out how living foraminiferal assemblages changed since the 1970s, we recalculated the percentages of species, observed as living in 1970, from the archive material stored at the Institute of Geosciences (University of Kiel). Unfortunately, only two of Exon's (1972) stations, 525-1 and 528-4, can be compared to our data because they were located next to our stations PF16-25 and PF16-28.

According to Exon (1972), a southern Gelting Bay profile extended from 8 m to 22 m and did not exhibit any living individuals of foraminifera at the deepest station. At the shallow water station

525-1, an arenaceous *Ammotium cassis* was found as a dominant species (83.9 %) and all other species (including *Ammonia beccarii*) were rare. Conversely, our survey showed a clear dominance of *Elphidium albiumbilicatum* (56 %) and *A. beccarii* (25 %) at station PF16-25; *Elphidium excavatum* subspecies were common (17 %); *E. incertum* and *A. cassis* were rare (1 %).

In the outer Flensburg Fjord, the only station with a predominance of *Ammotium cassis* (96%) and rare occurrence of *Ammonia beccarii* (1 %) was site 528-12, whereas at all other stations *Elphidium incertum* prevailed (Exon, 1972). At station 528-4, *E. incertum* was dominant and *E. excavatum* was common. As compared to the 1970s, our station PF16-28 (Appendix 2) showed a clear dominance of *E. excavatum* subspecies (together 87 %) with *A. beccarii* as common (6 %) and other species (*E. incertum, E. albiumbilicatum, E gunteri* and *A. cassis*) as rare (maximum 4 %).

2.5.6 Changes in species composition

The usage of the total fauna (living plus dead) makes comparisons of species composition from different studies difficult due to post-mortem changes in a population. This relates in particular to test destruction and transport (Murray, 1989; 2006). Almost all of Rottgardt (1952) stations, considered in this study, are situated within sheltered depositional areas of Flensburg Fjord, where muddy sediments prevailed. Furthermore, the Baltic Sea is undersaturated with respect to calcium carbonate (Jarke, 1961; Grobe & Fütterer, 1981) and calcareous tests are dissolved within a few weeks after death (Hermelin, 1987). The high abundances of calcareous foraminifera in total assemblages indicate either the high productivity of certain species, or their dominance in the assemblage while they were alive.

In this study, we observed a decline of *Ammotium cassis* and a flourishing of *Ammonia beccarii* in shallow areas of Gelting Bay, as compared to the previous study of Exon (1972). Only isolated individuals of *A. cassis* were found in Gelting Bay and in the outer fjord. The extraordinary high abundances of *A. cassis* in the 1970s in the uppermost centimetre of sediments reflected a situation lasting at least 3.7 years (from 1966 to 1970), if we apply a sedimentation rate in Gelting Bay of 2.7 mm/year (Müller *et al.* 1980; Fig. 2). Indeed, changes in species composition since the 1970s are similar to those we observed in Kiel Fjord (Nikulina *et al.*, 2008), where living *A. cassis* was abundant in the 1960s (Lutze, 1965) and is absent today.

In shallow areas of Gelting Bay, it was reported that the widespread colonization of *Ammonia beccarii* takes place above the discontinuity layer but never below it (Exon, 1972). In its turn, the maximum abundances of *Ammotium cassis* were found in association with the presence of a halocline, which provides a high input of particulate organic matter (Olsson, 1976), or within sediments where the redox boundary sustains high bacterial numbers (Linke & Lutze, 1993). The introduction of this species into the Baltic Sea took place between 1935 and 1952 (Lutze, 1965), and it was frequent in Kiel Bight between the early 1960s and mid 1980s (Schönfeld & Numberger, 2007a). In Flensburg Fjord, *Ammotium cassis* was not reported by Rottgardt (1952) in the late 1940s. It is likely that due to the lower frequency, or intensity of salt-water inflows from the North Sea in the past decades (Matthäus, 2006), the establishment of a stable discontinuity layer, a necessary condition for nutrition of *Ammotium cassis*, was impeded, and therefore this species became extremely rare in the outer Flensburg Fjord. This pattern was previously shown for Eckenförde Bay (Schönfeld & Numberger, 2007a) and Kiel Fjord (Nikulina et al., 2008). Taking into account a preference of *Ammotium cassis* to dwell close to discontinuity layers in the sediment column (redox boundary) and at the sediment-water interface (halocline), we conclude that both niches are nowadays occupied by more tolerant and opportunistic *Elphidium incertum* in the inner fjord and *Ammonia beccarii* in Gelting Bay.

At the same time, isolated specimens of *Ammotium cassis*, observed in the outer parts of Flensburg Fjord in 2006 require explanation. Do they represent a relict population, survivors of the periods of lower salinity, or are the forerunners of the reintroduction of this species? To answer these questions, further monitoring of foraminiferal assemblages in the Kiel Bight, and, in particular, Flensburg Fjord is needed.

2.5.7 Absence of Eggerelloides scaber

Eggerelloides scaber was reported by Lutze et al. (1983) as a species whose distribution does not depend directly on the substrate type, or bottom topography in the SW Baltic. This species does require salinity conditions of at least 24 psu lasting most of the year. Rottgardt (1952) reported high abundance of *Eggereloides scaber* in samples taken in the outer Flensburg Fjord. In 1960-70s, *Eggerelloides* was virtually absent in the outer fjord (Exon, 1972, Lutze et al., 1983), and we also did not observe this species in 2006. The absence of living *Eggereloides scaber* in our study apparently reflects the reduction of salt-rich Kattegat water inflows to the Western Baltic, as similarly shown by Lutze et al. (1983) for marginal basins of Kiel Bight.

2.6 CONCLUSIONS

Comparison with previous studies from late 1940s and 1970s revealed apparent changes in species composition in the outer Flensburg Fjord: a decline of arenaceous *Ammotium cassis*, a flourishing of calcareous *Ammonia beccarii* in Gelting Bay and a dominance of *Elphidium incertum* in the inner fjord. These changes are similar to those reported recently from other fjords of the SW Baltic Sea (Eckenförde Bay and Kiel Fjord) and are most likely associated with the generally decreased intensity and frequency of major Baltic inflows since 1960s caused by larger freshwater supply from the catchment area and changes in atmospheric circulation over the north Atlantic during the past decades (Meier et al. 2006).

Five foraminiferal biofacies were distinguished in Flensburg Fjord. Their distribution appeared to be controlled mainly by sediment grain size and food availability, while oxygenation of bottom waters was not a limiting factor for foraminifera. Correlations of some species to grain-size and food particles support the idea about role of grain size and food in distribution of benthic foraminifera in the Flensburg Fjord. The inner Flensburg Fjord (Biofacies 1) was dominated by *Elphidium incertum* dwelling within muddy sediments rich in organic matter. Biofacies 2 comprised the "*E. incertum – E. excavatum*" group, which was found in the muds and sandy-muds of the fjord loop around Holnis Peninsula and the outer fjord with less food availability. Gelting Bay reflects a distinctly different area with a shallow-water, brackish and sandy habitat poor in food particles. This area comprises the assemblage of Biofacies 3 dominated by opportunistic *Ammonia beccarii* and *Elphidium albiumbilicatum*. *A. beccarii* and *E. incertum* were dominant in the central Flensburg Fjord and near-shore zones of the loop around the Holnis Peninsula (Biofacies 4) with sandy muds relatively poor in food particles. *Elphidium excavatum* subspecies inhabited the innermost part of the fjord (Biofacies 5) with the fine-grained muddy sediments rich in food.

It was suggested that the frequent occurrence of the infaunal *Elphidium incertum* in the uppermost sediment layer might reflect the seasonal hypoxic-anoxic conditions in the sediment column of the inner Flensburg Fjord, which preceded the sampling period. It appears, however, that an A/E index, an indirect hypoxia proxy, is not applicable for the SW Baltic Sea.

ACKNOWLEDGEMENTS

The authors are grateful to crew of RV Polarfuchs Holger Meyer and Helmut Schramm for help with sampling. We acknowledge Udo Laurer (IFM-GEOMAR, Kiel), Ute Schuldt, Brigitte Salomon (Institute of Geosciences, Kiel) and Irina Gembitskaya (St. Petersburg State Mining Institute, St. Petersburg) for technical assistance. Also we are indebted to Wolfgang Kuhnt (CAU, Kiel), who provided the access to archives of Institute of Geosciences and made Exon's archive data available. We also appreciate the constructive comments and valuable suggestions of J.W.Murray and L. de Nooijer, who reviewed the earlier version of the manuscript. This study was financed by Leibniz Award DFG DU 129-33 and a Scholarship of German Academical Exchange Service (DAAD).

Chapter 3

Test abnormalities of recent benthic foraminifera in the western Baltic Sea

Published as:
Polovodova, I., and J. Schönfeld., 2008. Foraminiferal test abnormalities in the western Baltic Sea. *Journal of Foraminiferal Research*, 38 (4): 318-336

ABSTRACT

Abnormal tests were commonly found in recent benthic foraminiferal assemblages in the Kiel Bay, the western Baltic Sea. We assessed 18 different types of abnormalities, which were classified into five groups: chamber, apertural-, umbilical-, coiling and test abnormalities. In both fjords, test abnormalities are over-represented in *Ammonia beccarii* and under-represented in *Elphidium excavatum* subspecies compared to their average proportions in the living assemblages. We found two species-specific abnormality types (a bulla-like chamber covering the umbilicus and spiroconvex tests), which occurred only in *Ammonia beccarii*.

In the outer Kiel and Flensburg fjords, the highest frequencies of abnormal tests were associated with occasional salt-rich bottom-water inflows from the Belt Sea. Based on the predominance of megalospheric specimens of living foraminifera, it is suggested that coincidence of salinity changes with a reproduction period might be harmful, especially for young individuals, leading to development of abnormal tests. On the other hand, pollution by heavy metals led to higher percentages of abnormal tests in the inner parts of both fjords. Our data show different relationships of abnormal tests with heavy metals in both fjords due to different hydrographical conditions.

Tests of *Ammonia beccarii* found in Gelting Bay, Flensburg Fjord, showed traces of dissolution and development of double tests. Such specific abnormal tests mirror the peculiar environmental setting characterized by changes in salinity and enhanced sediment redeposition. It is concluded that abnormal tests as an indicator of environmental pollution have to be used cautiously in areas with strong environmental instability.

3.1 INTRODUCTION

A growing number of studies have reported morphological abnormalities of foraminiferal tests from marine (among others, Watkins, 1961; Alve, 1991; Boltovskoy and others, 1991; Sharifi and others, 1991; Yanko and others, 1994, 1998, 1999; Geslin and others, 2002; Bergin and others, 2006) and experimental settings (Mikhalevich, 1976; Wennrich and others, 2007).

Deformed foraminiferal tests are considered to come from disruption of the growth plane leading to an abnormal shape, which is different in comparison with other specimens of the same species (Murray, 2006). A set of experiments was conducted in order to find out the factors responsible for formation of abnormal tests. Conditions of unfavourable salinity (Stouff and others, 1999 a, b), acidification (Le Cadre and others, 2003), food supply (Murray, 1963), and elevated levels of heavy metals (Sharifi and others, 1991; Saraswat and others, 2004; Le Cadre and Debenay, 2006) were the main factors inducing test abnormalities. In addition, a high proportion of abnormal tests may be induced by intense hydrodynamic conditions (Geslin and others, 2002).

Recently, more attention has been focused on anthropogenic origin of malformations. Several authors (e.g., Alve, 1991; Sharifi and others, 1991; Yanko and others, 1994, 1998; Samir and El Din, 2001; Bergin and others, 2006; Nigam and others, 2006; Di Leonardo and others, 2007) reported increased frequencies of abnormal foraminiferal tests in estuaries subjected to heavy metal pollution. Moreover, Sharifi and others (1991) and Samir and El Din (2001) showed that deformed tests contained a higher proportion of heavy metals, such as Pb, Zn, Cu, Cr and Cd, than normal ones. Ernst and others (2006) noted the development of abnormal tests in a mesocosm experiment of an oil spill.

Previous studies have described a wide spectrum of malformation types. For instance, Alve (1991) distinguished seven types of morphological deformities: aberrant chamber shape and size, twisted or distorted chamber arrangement, protuberances, multiple apertures, enlarged apertures, reduced chamber size and twinned forms. On the other hand, Yanko and others (1998) distinguished 11 abnormality types of which wrong coiling, poor development of the last whorl, additional chambers, irregular keel, lateral asymmetry and lack of sculpture added to the list of Alve (1991). In addition, over-developed chambers, an excessively high spiral side (spiroconvex tests), a bulla-like chamber covering the umbilicus, an umbilical plug and inflated or deflated tests were reported by Sharifi and others (1991), Samir and El Din (2001), Bergin and others (2006) and Ernst and others (2006) respectively.

Several attempts to classify or systematize abnormal tests have been made. Boltovskoy and

Chapter 3 – Test abnormalities

Wright (1976) categorized foraminiferal abnormal tests according to their origin: physically (mechanically) or ecologically induced. Yanko and others (1998) subdivided abnormalities into early, intermediate and adult stages, according to the number of whorls and chambers. A further distinction was suggested by Stouff and others (1999b), who stated that *malformations* are abnormalities that take place during ontogenetic development, whereas *deformities* occur during the life of the adult foraminifer and proposed the term of *morphological abnormalities* when the origin of abnormality was not evident. In the following we will use the terms deformities and morphological abnormalities.

This study aims to (i) describe the distribution of abnormal tests in Kiel and Flensburg fjords of Kiel Bay; (ii) identify the different types of test abnormality; (iii) classify abnormalities according to morphological criteria and (iv) clarify the factors responsible for the formation of abnormal foraminiferal tests.

3.2 REGIONAL SETTING

The Kiel Fjord (54°19' - 54°30' N; 10°06' - 10°22' E) is a narrow 9.5-km long inlet, south-western Baltic Sea (Fig. 3.1). It comprises two basins: the southern inner fjord (up to 250 m wide) and the northern outer fjord (up to 7.5 km wide), which merges into the Kiel Bay. There is a network of channel-like depressions, which connect the inner and outer fjords, beginning in the southern part at 14 m water depth and then sloping into the Kiel Bay at approximately 20 m water depth.

Water masses of the Kiel Fjord are similar to those in the Kiel Bay. During summer, the water column is well stratified: surface water of 16 °C and average salinity of 14 practical salinity units (psu) overlays deep water of 12 °C and up to 21 psu. In winter and spring, the stratification is less pronounced, and water masses of the inner fjord of 2 °C are uniformly mixed (Themann, 2002). The influence of occasional salt-rich bottom water inflows from the Belt Sea with a salinity of up to 33 psu, apparently does not play a significant role for the hydrography of the inner Kiel Fjord (Fennel, 1996).

The most important source of sediment to the Kiel Fjord is Pleistocene till which is eroded from cliffs and shoals in north-western part. Shallow coastal areas are characterized by lag sediments with coarse sand and gravel, which grade into sandy muds and silts in depressions. In the innermost fjord, dark organic-rich muds are encountered even in shallow areas. Foraminiferal tests and shells of other carbonate-producing organisms are subjected to abrasion and corrosion in the sediments. Abrasion and redeposition processes play an important role in the shallow areas of the Kiel Bay, whereas corrosion of foraminiferal tests takes place in the deeper basins due to undersaturation of carbonate in bottom water (Grobe and Fütterer, 1981).

The Kiel Fjord area is a highly urbanized. The town infrastructure, numerous shipyards, military and sport harbours and the intensive traffic through the Kiel Canal caused an increase of anthropogenic impact over the last 70 years. In particular, the shipbuilding industry leads to substantial heavy metal, oil and TBT (tributyltin) pollution (Helland and Bakke, 2002).

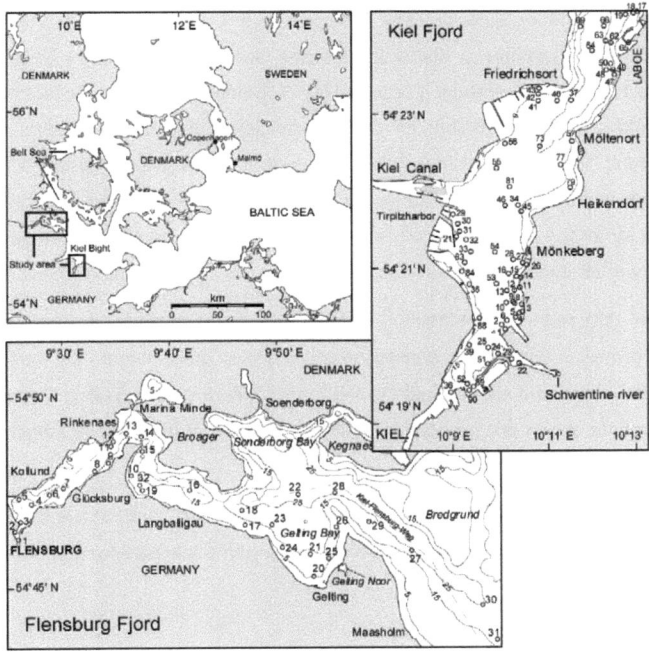

Figure. 3.1: Study area. Black circles indicate here the sampling points. Numbers of stations are given here without prefixes PF15- for the Kiel Fjord and PF16- for the Flensburg Fjord for sake of convenience.

Flensburg Fjord (53°41' - 55°00' N; 9°24' - 10°10'E) is a narrow, 50-km long, east-west trending inlet in the north-western part of the Kiel Bay. The Flensburg Fjord is subdivided into the inner fjord (10-20 m deep; 1.3-3 km wide) and the outer fjord (10-32 m deep; 4 km wide) by the Holnis Haken Shoal, situated off the Holnis Peninsula. The outer Flensburg Fjord comprises the 13-31 m deep Sonderborg Bay, Gelting Bay (4-22 m deep) and open waters east of the Gelting

Peninsula with a very high depth range – from 5 m in the area of Schleisand up to 39 m in the Little Belt. The Gelting Noor is a partly enclosed estuary of the Gelting Brook.

Water exchange with the Kiel Bay is distinctly better in the outer Flensburg Fjord than in the inner part. Flensburg Fjord is the most protected estuary in the region, and wave action does not reach significant depth, thereby sediment zones are shifted towards shallower waters. During the winter, the inner fjord is well mixed at 6.5 °C and a salinity of 21 psu. The water column is well stratified during the summer: surface water averages 17.5 °C and 16.5 psu and bottom water is 11 °C and 18 psu. Every summer, a stable thermocline develops in the inner fjord at 8-9 m depth (Gemeinsames Komitee Flensburger Förde, 1974). Together with enhanced eutrophication in the 1980's, this setting was responsible for oxygen deficiency that lasted several months without interruption (Wahl, 1985). In the outer fjord, a persistent pycnocline at 16-20 m (Schwarzer and Themann, 2003) separates brackish surface water from salty deep water throughout the year. In outer fjord, the top of the pycnocline coincides with the effective depth of wave action and divides the sedimentary environments into non-depositional and depositional areas (Exon, 1972).

In general, the sediment distribution in Flensburg Fjord is similar to that in the Kiel Fjord. Coarse sands prevail in the shallow coastal areas, and they grade into sandy muds and silt in the deep basins. Dark mud and silt dominate in the inner shallow areas. Gelting Bay is characterized by sandy sediments, which are predominantly transported from the east by longshore drift (Exon, 1971).

The Flensburg Fjord was strongly exposed to fertilizers and sewage outfalls from the adjacent land in the 1980's. Presently, it is a resort area with a few harbours, yacht traffic and small towns on the shore.

3.3 MATERIAL AND METHODS

Surface sediment samples were taken on 10 daily cruises of the R/V *Polarfuchs* in the area of the Kiel Fjord between December 2005 and May 2006, and in Flensburg Fjord in June 2006. The majority of samples were retrieved with a Rumohr corer, which has a sampling tube of 56 mm inner diameter. A van Veen grab was used to recover sandy sediments.

Within minutes after sample retrieval with Rumohr corer, salinity, temperature and dissolved oxygen content of bottom waters were measured on board with oxygen- and conductivity meters (WTW Oxi323/325 and LF320, respectively). The uppermost 1 cm of the sediment was scraped off with a spoon. When sampling with a van Veen grab, cut-off syringes marked with a centimetre-scale were used for sampling. The sample was transferred into a glass vial, homogenized and subsampled for TOC, TN, SiO_2, Chl a and heavy metals (Cu, Zn, Sn and Pb) analyses. The remaining sediment was transferred to an acid washed PVC vial, then preserved and stained with a rose Bengal and ethanol solution (2 g/l) (Murray and Bowser, 2000). In total, 77 and 32 samples were taken in Kiel and Flensburg Fjords respectively (Fig. 3.1).

The samples were first passed through a 2000-μm screen in order to remove mollusc shells and pebbles, and, then they were gently washed with tap water through a sieve with 63-μm openings. These fractions (63-2000 μm and >2000 μm) were dried at 60 °C and weighed. The 63-2000-μm fraction was then further split. In order to assess the response of foraminifera to the recent environmental changes reflected in occurrence of test abnormalities in both fjords, only living foraminifera were studied. All stained foraminifera, considered as living at the time of sampling, were picked out from respective aliquots, sorted at the species level, mounted in Plummer cell slides with glue and counted. Abnormal foraminiferal tests were counted and different types of abnormalities were determined and counted separately. The main types of abnormal tests were photographed using a scanning electronic microscopes (SEM), a JSM-6460LV (St. Petersburg State Mining Institute) and a Cam Scan-CS-44 (Institute of Geosciences, Kiel University). Light micrographs of foraminiferal tests were taken with an Olympus MIC-D digital microscope.

To qualitatively estimate the heavy metals within the foraminiferal tests, analyses were made on normal and abnormal tests using energy dispersive spectrometer (EDS) attached to SEM (Cam Scan-CS-44). X-ray spectra were obtained at 20 kV accelerating potential and measured in counts (live counting time range 36-77 s). Due to the heterogeneous distribution of trace metals (Severin, 1990), at least three points on each test were measured to check for internal variability of the shell composition.

To measure geochemical parameters in sediments, samples were prior freeze-dried and powdered in an agate mortar. Measurements of total organic carbon (TOC) and total nitrogen (TN) were performed with a Carlo Erba NA-1500-CNS analyzer with accuracy better than ±1.5%. Chlorophyll a was determined after acetone extraction with a Turner TD-700 Fluorometer with a precision of ±10%. Biogenic silica (SiO_2) measurements were done using a Skalar 6000 photometer with precision ±1%. For heavy metal analyses, the bulk sediment was digested in a HNO_3-HF-$HClO_4$-HCl mixture solution and measurements of total concentrations of Cu, Zn, Pb and Sn were performed with an AGILENT 7500cs ICP-MS. Blanks and standard MAG-1 were repeatedly analysed together with the samples in order to evaluate the precision and accuracy of the measurements. The accuracy of analytical results as estimated from replicate standard measurements was better than ±1.5%.

A detailed review of sediment geochemistry, including analytical methods can be found in Nikulina and others (2007). Results of geochemical characteristics of surface sediments in the Flensburg Fjord will be given elsewhere.

For correlation analysis of geochemical parameters and foraminiferal data Pearson's correlation coefficient was used. Statistical analysis was performed by means of software *STATISTICA 6.0*.

3.4 RESULTS

3.4.1 Hydrography

Since our sampling campaign in Kiel Fjord comprised several seasons, the temperature and salinity of near-bottom water showed a pronounced seasonality. Temperature decreased from 8°C on average in December 2005 to 2°C in February, and raised again to 7°C in May 2006. In December 2005, the near-bottom water showed the highest salinity with 23.2 psu and minimum values of 16.5 psu in May. In the Schwentine river mouth, the boundary layer between riverine fresh water and saline fjord water was encountered at approximately 1 m depth in February.

The oxygen concentration mostly exceeded 400 μmol/l and decreased slightly only in the deep basins. The saturation levels varied from 58 % to 100 %. As such, a sincere oxygen deficiency in the near-bottom waters of Kiel Fjord was not recognized.

In Flensburg Fjord, temperature and salinity of near-bottom water varied in range 7.2 -13.7°C and 18.3 - 25.4 psu, respectively, in June 2006. The lowest temperature and highest salinity were encountered in depressions of the inner and outer parts. The highest temperature in Flensburg Fjord we observed in southernmost Gelting Bay. Two stations in Gelting Bay (PF15-20 and PF15-26) showed the lowest salinity values (18.3 and 18.9 psu), which reflect, apparently, the influence of Gelting Brook, bringing fresh-water to this area.

The content of dissolved oxygen in near-bottom water of Flensburg Fjord was lower than in Kiel Fjord and ranged from 160 to 308 μmol/l with the highest value off Gelting Noor. The saturation levels varied from 48 to 100%. We also observed no oxygen deficiency in this area.

3.4.2 Definition of abnormality types

Seventeen types of aberrant foraminiferal tests were recognized in the Kiel Fjord and 15 types in Flensburg Fjord. All abnormal tests were divided into five groups:

1) *Chamber abnormalities*, which include aberrant chamber shape (Pl. 4, Fig. 6); twisted chamber arrangement; additional chambers (Pl. 4, Figs. 5, 8); reduced chamber size (Pl. 4, Fig. 4, 13, 22; Pl. 5, Fig. 14); overdeveloped chambers of the last whorl (Pl. 4, Fig. 14, 21) and protuberances. Protuberances and abnormally protruding chambers were previously described by Alve (1991), Almogi-Labin and others (1992),Yanko and others (1994), Geslin and others (2000; 2002). Identification of these two abnormalities is a challenging task, because they may be confused with the frustrated double tests (Stouff and others, 1999a). During the early ontogenetic stage, two second chambers may grow from one proloculus with subsequent development of an independent whorl from each of the second chambers. If only one whorl develops from one of these chambers, then the other appears as a protuberance on the proloculus (Stouff and others, 1999a: Pl. 2, Fig. 4).

2) *Apertural abnormalities,* which consist of multiple apertures on tests (Pl. 4, Fig. 15; Pl. 5, Fig. 12).

3) *Abnormalities of the umbilical side* of the test are represented by a bulla-like chamber covering the umbilicus. This feature was previously described by Frontalini and Coccioni (2008: Pl. 2, Fig. 9) as an abnormally protruding chamber, whereas Bergin and others (2006) reported it as umbilical plug. Unfortunately, the latter did not provide images of their abnormal tests. As such, we can only suggest that the abnormal umbilical plug, mentioned by Bergin and others (2006), differs from the normal plug of schizont test morphology, reported by Stouff and others (1999c).

4) *Abnormal coiling* includes a wrong direction of coiling, poor development of the last whorl (Pl. 4, Fig. 24), and development of several whorls with different axes of rotation (Pl. 4, Fig. 16; Pl. 5, Fig. 7 a-b, 8, 13).

5) *Test abnormalities* comprise several structures including an irregular keel, twinning (Pl. 4, Fig. 9; Pl. 5, Fig.4), lack of sculpture (Pl. 4, Fig. 12), an excessively high spiral side or spiroconvex tests (Pl. 4, Fig. 7), twisting of entire test (Pl. 5, Fig. 9, 15), compressed tests (Pl. 4, Fig. 19; Pl. 5, Fig. 3) and non-developed tests. The latter represent tests with smoothed ornamentation, but with distinguishable chambers, as compared to the tests with lack of sculpture.

Chapter 3 – Test abnormalities

Some tests showed multiple apertures and several whorls with different axes of rotation. It is likely that those may correspond to the double and multiple tests (Stouff and others, 1999a), which were described by some authors also as twinned tests (Alve, 1991; Sharifi and others, 1991; Yanko and others, 1998).

3.4.3 Kiel Fjord

Distribution of abnormal tests

The living benthic foraminiferal fauna in Kiel Fjord includes nine species, among which *Ammonia beccarii* (52% of all specimens on average) and subspecies of *Elphidium excavatum* (44% on average) are dominant (Appendix 4). *E. incertum, E. gerthi, E. albiumbilicatum, E. williamsoni, E. guntheri, Ammotium cassis* and *Reophax dentaliniformis* are rare (less than 3%).

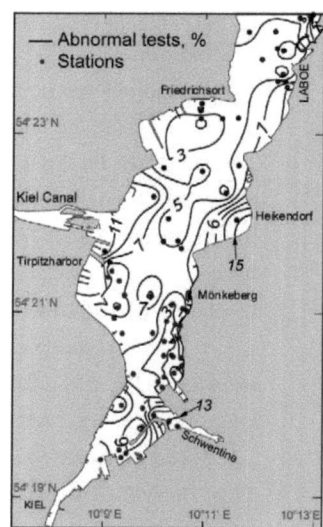

Figure 3.2: Spatial distribution of tests with abnormalities in Kiel Fjord.

The morphological abnormalities were encountered in six species (App. 4.1). The majority of abnormal tests were found in *A. beccarii*, which have 73% of all abnormal tests observed in living specimens in this area. Less abundant were abnormalities in *E. excavatum excavatum* (16%), *E. excavatum clavatum* (4%), *E. gerthi* (3%), *E. incertum* (3%) and *E. albiumbilicatum* (1%). It is evident from the percentages that abnormalities are over-represented in *A. beccarii* and under-represented in *E. excavatum* subspecies compared to their average proportions in the living assemblages of Kiel Fjord.

The percentage of abnormal tests recorded in samples collected in the Kiel Fjord range between 0 - 25%. Abnormality frequencies higher than 10% were found in ten of 77 samples. High proportions of abnormal tests occurred in samples from marginal sites (Fig. 3.2) to the north offshore from the Laboe resort (PF15-17; 25%) and in the coastal zone near Heikendorf (17%). A transect consisting of four stations (PF15-22 – PF15-25) taken from the inner part of Schwentine river through its estuary to the western shore of the fjord demonstrated higher percentages of abnormal tests close to the river mouth (15%) compared to the western side (11%). Relatively high proportions of abnormal tests were also detected near the Tirpitz Harbour (13%), which is used by the military, near a former oil pier (12%) in Mönkeberg, and in the central and north-western parts of the inner fjord (10%). Stations PF15-20, PF15-60 and PF15-61 (Fig. 3.1) in the outer Kiel Fjord showed relatively low proportions of abnormal tests (4% on average).

Diversity of abnormal tests

The maximum variety of test abnormalities was seen in the most abundant taxa, *Ammonia beccarii* and *Elphidium excavatum* subspecies. Only three types of deformities were observed in *E. albiumbilicatum*, which inhabits an area with strong currents in the Kiel Fjord (Nikulina and others, 2007).

Specimens of *Ammonia beccarii* showed the next dominant abnormalities: reduced size of chambers (22% of all abnormalities in this species), aberrant chamber shape (19%), overdeveloped chambers (14%), excessively spiroconvex tests (14%) and compressed tests (8%). The prevailing abnormalities of *Elphidium excavatum excavatum* were reduced chamber size (36%), aberrant chamber shape (20%) and compressed tests (17%). Fifty percent of the abnormal tests of *E. excavatum clavatum* had reduced chambers, 18% were compressed and 10% showed the development of two whorls with different axes of rotation. Reduced chamber size prevailed also in *E. incertum* and *E. gerthi* (App. 4.1). Only five abnormal tests of *E. albiumbilicatum* were found, and three of them showed overdeveloped chambers. It comes out that, with the exception of *E. albiumbilicatum,* all *Elphidium* species have a similar diversity of abnormality types.

The lack of sculpture and the occurrence of bulla-like chamber covering the umbilicus were rare in the Kiel Fjord. However, a bulla-like chamber and excessively spiroconvex tests are common only to *Ammonia beccarii*, due to its low trochospiral test morphology.

Heavily deformed foraminiferal tests were also encountered mainly in *A. beccarii*. These tests showed several types of abnormality, sometimes making their taxonomical identification difficult (Pl. 4, Fig. 23). We considered such tests as a complex form of test abnormalities (App. 4.1).

3.4.4 Flensburg Fjord

Distribution of abnormal tests

The living benthic foraminiferal assemblages in Flensburg Fjord are dominated by *Elphidium excavatum* subspecies (37% on average), *E. incertum* (28% on average) and *Ammonia beccarii*

Chapter 3 – Test abnormalities

(25% on average). *E. albiumbilicatum* is common (10% on average), whereas *Reophax dentaliniformis regularis, Ammotium cassis, E. gerthi* and *E. williamsoni* are rare (1% on average).

The proportion of specimens with aberrant tests varies between 0 - 19% in Flensburg Fjord. Most of the abnormal individuals come from the dominant species: *Ammonia beccarii, Elphidium incertum* and *E. excavatum* subspecies (representing 40, 32 and 25% of all deformed tests, respectively). This shows again that tests abnormalities are over-represented in *A. beccarii* (with reference to the average proportion of this species in the living assemblages of the Flensburg Fjord).

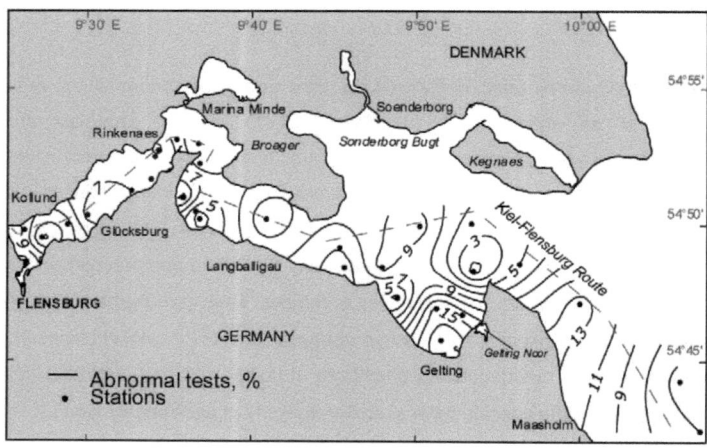

Figure 3.3: Spatial distribution of abnormal test frequencies in the Flensburg Fjord.

The highest proportion of abnormal specimens (19%) was recorded in the Gelting Bay (Fig. 3.3). Furthermore, we observed high abundances of abnormalities in the innermost (PF16-04) and outer (PF16-10 and PF16-27) fjords. More than 10% of abnormal tests were seen at four stations that are situated along the shipping lane "Kiel-Flensburg Route" (Fig. 3.3). Low percentages of abnormalities were encountered in the eastern Gelting Bay (3%) and at station PF16-19 (1.8%) in the outer fjord.

Diversity of abnormal tests

The highest variety of abnormalities in the Flensburg Fjord was observed in *Ammonia beccarii* and *Elphidium incertum*, which respectively showed 12 and 11 abnormality types (App. 4.2). Ten and five different types of abnormalities were observed in the tests of *E. excavatum* subspecies and *E.*

albiumbilicatum, respectively.

Ammonia beccarii showed excessively spiroconvex tests (26%), reduced chambers (22%) and aberrant chamber shape (16%) as dominant abnormalities. The most of abnormal tests in *Elphidium excavatum excavatum* had reduced chambers (39%). The latter prevailed also in *E. excavatum clavatum* (36%), *E. albiumbilicatum* (five of ten tests) and *E. incertum* (25%). These species also showed aberrant chamber shape (22%) as a common type of abnormal tests. Thus, the distribution of the abnormality types in *A. beccarii* is different between Flensburg Fjord and the Kiel Fjord, whereas *Elphidium* species showed similar abnormal test types between these areas.

Multiple apertures, irregular keel and poor development of the last whorl are rare types of foraminiferal test abnormalities in the Flensburg Fjord. Only two tests exhibiting several types of abnormality were found in *A. beccarii* and *E. incertum*.

In the eastern Gelting Bay, near Gelting Noor, we found specimens of *Ammonia beccarii* with unusually thin or opaque shell walls (Pl. 6, Fig. 2, 3), and extremely corroded tests (Pl. 6, Fig. 4, 5). In some cases, only the inner organic lining was left (Pl. 6, Fig. 6). Some of these specimens also showed a distinct type of abnormality, where a small, deformed foraminifer was firmly attached to a bigger corroded or destroyed test (Pl. 7, Figs. 4, 9). The chambers of the smaller specimen were deformed; more tightly arranged and had extremely thin walls. Samples taken in the southern and central Gelting Bay also contain tests of *A. beccarii* with slight traces of corrosion or dissolution. All corroded specimens of *A. beccarii* in the Gelting Bay showed a bright rose-Bengal staining and were considered alive when sampled.

3.4.5 Distribution of abnormalities between living and dead assemblages

To test fossil assemblages for abnormal tests, we analyzed the total foraminiferal fauna in one control sample, taken in the outer Kiel Fjord in 2007 at the same location, where sample PF15-38 was taken. As a result, we observed 26% of abnormal foraminiferal tests in the uppermost centimetre of sediments, which comprises approximately 10 years. Keeping in mind the value of 4% observed among living specimens at station PF15-38, we may assess how many abnormal specimens were present in dead fauna (22%). A distinct difference (18%) in abnormalities between living and dead assemblages indicates that elevated frequencies of abnormal tests are phenomenon, which takes place not only recently, but was also present in Kiel Fjord over the past decades, when environmental stress and anthropogenic pollution were higher.

Distribution of abnormalities between different species in dead assemblage was similar to that in living fauna. In both cases abnormal specimens were over-represented in *Elphidium excavatum* subspecies. Also we found a predominance of reduced (44%) and overdeveloped chambers (33%) in dead assemblage. Reduced chamber size (29%) was also one of the most abundant abnormality types in living fauna, together with aberrant shape (43%) and twisted arrangement (29%) of the chambers.

3.4.6 Test abnormalities and sediment geochemistry

In order to reveal the response of benthic foraminifera to recent human-induced stress in Kiel and Flensburg Fjords, surface sediment samples were analysed for a set of trace metals: Cu, Sn, Zn and Pb, which are associated with the shipbuilding industry. Copper, tin and zinc have been used, as biocides in antifouling paints (Bellinger and Benham, 1978; Clark and others, 1988; Helland and Bakke, 2002), whereas lead is known to come from boat and ship exhaust systems (Abu-Hilal and Badran, 1990) and also forms the pigment basis of anticorrosives and primer paints (V.-Balogh, 1988).

In Kiel Fjord, we found higher proportions of abnormal tests (17%) among living specimens at station PF15-79 off Heikendorf with exceptionally high lead content (2169 mg/kg), which is two orders of magnitude higher than 20 mg/kg, an average lead concentration known for nearshore muds (Chester and Aston, 1976). Likewise, a relatively high percentage of abnormalities (10%) in the central part of the fjord coincide with the high level of Sn (15 mg/kg), which is seven times higher than 2 mg/kg, an average concentration of tin found in nearshore muds (Wedepohl, 1960). In spite of this, no significant correlations between the total proportion of abnormal tests and content of heavy metals in the sediments were recognized in both fjords, (Fig. 3.4a, b). However, the limited number of samples where certain abnormalities were found and where heavy metals were measured impeded determination of the relationships between different types of abnormality and environmental parameters. Though, the proportion of tests with additional chambers (var = 121.36) in Kiel Fjord correlated negatively with sedimentary Sn ($r = -0.501$), this relationship failed Student t-test for significance (App. 4.3, Fig. 3.4c). On the other hand, the occurrence of tests with a wrong coiling showed significant negative correlations with TOC, TN and Chl *a* ($r=-0.522$; -0.578; -0.763, respectively).

In Flensburg Fjord, counts of test with additional chambers and twisted tests were available together with content of heavy metals only in five samples (Fig. 3.4d). These samples showed significant positive correlations between the proportions of tests with additional chambers and Cu (r

= 0.886), Zn (r = 0.857), Sn (r = 0.831) and Pb (r = 0.847) content in sediments. A negative correlation of this abnormality with TN was also found, and twisting of the entire test (var = 18,544) correlated positively with all the heavy metals measured (App. 4.3). However, these latter two relationships failed the significance test. In addition, EDS analysis did not reveal high concentrations of heavy metals in abnormal tests, as compared to normal ones from either fjord.

3.5 DISCUSSION

3.5.1 Abnormal tests as indicators of heavy metal pollution?

Presence of abnormal tests in benthic foraminiferal communities was reported from various types of environments, in particular those polluted by heavy metals (Alve, 1991; Sharifi and others, 1991; Yanko and others, 1994; 1998; 1999; Samir and El Din, 2001; Geslin and others, 2002; Bergin and others, 2006; Burone and others, 2006; Di Leonardo and others, 2007; Frontalini and Coccioni, 2008, among others). It is important to distinguish what percentage of abnormal tests should be regarded as "normal" in order to compare with percentages in environments subjected to pollution. Alve (1991) found 1% abnormal foraminiferal tests in unstressed foraminiferal assemblages. The same value was also obtained in laboratory experiments (Stouff and others, 1999b).

In Kiel Fjord we observed high abnormalities at the station PF15-79 and in the central fjord in sediments highly enriched in lead and tin, respectively. The extraordinarily high Pb levels in the sediment may result from analytical error due to the absence of replicate samples. However, coal and ash residues, which were commonly found in sediments of Kiel Fjord, may also contribute to high lead and tin concentrations (Erlenkeuser and others, 1974; Reeder and others, 2006). On the other hand, Kiel Fjord represented a strategically key area during the Second World War (RIIA, 1990) and it well might be that corrosion of old ammunition buried in vicinity of the sampling site led to high metal concentrations. In addition, the muddy sediments encountered at station PF15-79 also would explain high levels of lead and tin due to a presence of an active depositional centre.

Muddy sediments and organic matter are substrates for adsorption of heavy metals (Tessier and Campbell, 1987) and, therefore, affect metal speciation, mobility and bioavailability in sediments. The most bioavailable are oxides and carbonates of lead (Reeder and others, 2006). Unfortunately, we did not do selective chemical extraction of heavy metals from sediments and therefore may only speculate about their speciation and bioavailability to foraminifera. Ankley and others (1994) discussed the role of iron sulphide complexes in reducing bioavailability of lead in sediments and pore waters. They noted seasonality in concentrations of these complexes in surface sediments with minimum in winter and maximum in late spring - early summer. However, sample PF15-79 was taken in Kiel Fjord at the beginning of May, when concentrations of iron sulphides were supposed to be high to make lead unavailable for foraminifera. Apparently, there are some other factors responsible for elevated foraminiferal abnormalities off Heikendorf.

In turn, the positive correlations of tests with additional chambers to Cu, Pb, Sn and Zn in Flensburg Fjord confirm the harmful influence of pollutants. However, one should keep in mind, that a small sample size (n=5) makes these results debatable. Nevertheless, tests with additional chambers are abundantly found in the inner Flensburg Fjord where muddy sediments and the highest levels of heavy metals were encountered (Fig. 3.4 a, b). On the other hand, every year, sediments of the inner Flensburg Fjord are exposed to seasonal oxygen deficiency (Wahl, 1985; LANU, 2007). This provides the suitable conditions for the formation of iron sulphide complexes, an important binding phase for metals in contrast to aerobic sediments, where iron sulphides are readily oxidized (Ankley and others, 1994). Thus, during changes from reducing to oxidizing conditions heavy metals may become bioavailable (Forester, 1993; Siegel, 2002; Dalloway, 2005), explaining therefore, a positive relationship between additional chambers and heavy metals in sediments of Flensburg Fjord.

Such surprisingly different relationships between tests with additional chambers and Sn content in the sediments of Kiel and Flensburg fjords (Fig. 3.4 c, d) arise a question about their reasons. It is evident from our observations and hydrographical data that salinity, on average, is higher in Flensburg Fjord than in Kiel Fjord. As such, it could well be that the bioavailability of tin to foraminifera is higher in Flensburg Fjord than in Kiel Fjord, due to different physico-chemical conditions: higher salinity and therefore more alkaline pH. With exception of copper and mercury there is a rapid decrease in the proportion of heavy metals bound to organic matter as the salinity increases (Mantoura and others, 1978; Siegel, 2002) due to competition for the humic ligands with calcium and magnesium, which are highly abundant in seawater. Rüdel (2003) reported an increased bioavailability of organotin compounds at more alkaline pH values. Taking into account that toxicity of metallic and inorganic tin is low and the major source for this metal in study area are antifouling paints, it may be hypothesized that we are dealing with organotin compounds, which are toxic not only to molluscs (LANU, 2001), but also to foraminifera (Gustafsson and others, 2000).

Chapter 3 – Test abnormalities

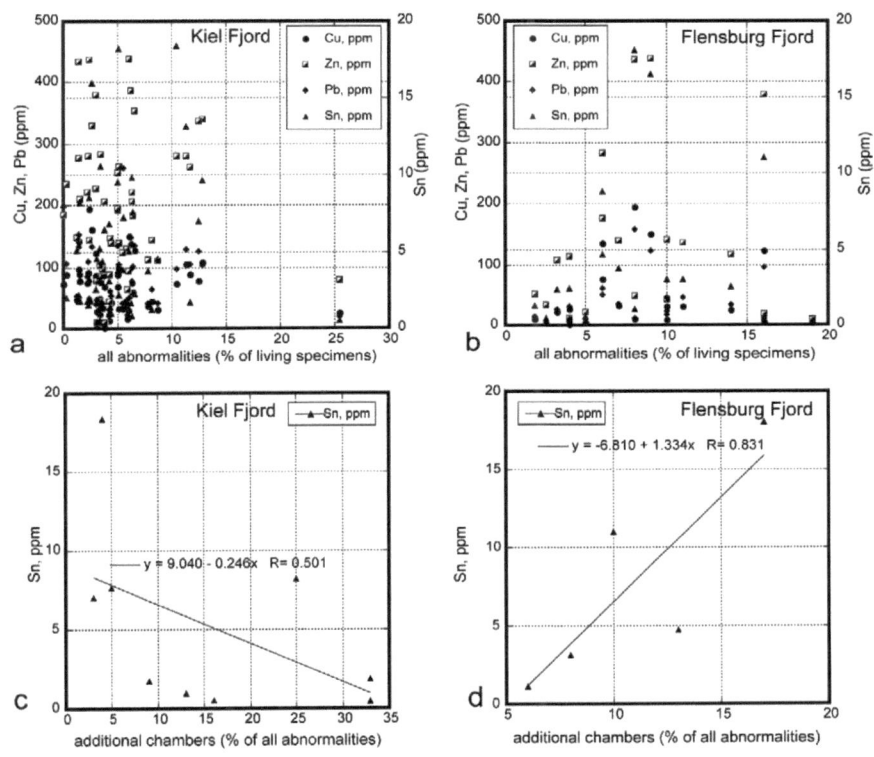

Figure 3.4: Relationships between test abnormalities and trace metals content of surface sediments for Kiel (a, c) and Flensburg (b, d) Fjords.

Alike in vertebrates the majority of absorbed lead ends up in bones and teeth (Barry, 1975), in foraminifera it should be apparently accumulated in carbonate test. It was shown that heavy metals might induce a crystalline disorganization and formation of interlammelar cavities, leading to test abnormalities (Geslin and others, 1998). To test these hypotheses, we did EDS analysis of heavy metals in foraminiferal tests. However, there were no differences in metal content between normal and abnormal tests. Our results are contradictory with those of Samir and El Din (2001), who reported the higher concentrations of Cu and Zn in abnormal specimens. This may be explained either by lower resolution of EDS analyzer or by less counting time (57 s on average) we used. Geslin and others (1998) also did not observe any differences in metal concentrations between normal and abnormal tests and suggested that metal analysis with ICPMS technique

Chapter 3 – Test abnormalities

would give better results than EDS.

An increase in the proportion of abnormal tests from the western part of the Kiel Fjord to the Schwentine river mouth can be related either to the high input of organic matter with river run-off or to changes in salinity. In the inner, middle and outer parts of the river the excessively spiroconvex tests of *Ammonia beccarii* were found. Spiroconvex tests were reported as an indicator of high organic-matter content (Seiglie, 1975; Samir and El Din, 2001). On the other hand, a shift from saline to freshwater conditions apparently is also unfavourable for benthic foraminifera. As suggested by Wennrich and others (2007), abnormal tests may also result from hyposalinity. Owing to the river discharge, the salinity is variable in the Schwentine area and may thus lead to the increased proportion of test abnormalities.

Excessively spiroconvex tests are also frequent in the inner (PF15-90) and central fjord (PF15-14; PF15-45) and off Heikendorf (PF15-79), where they coincide with high TOC content. An exception was at site PF15-17, which had a low content of all organic compounds and heavy metals, yet showed the highest proportion (25%) of total abnormal tests in all of the Kiel Fjord. Similar settings were observed in Flensburg Fjord, where the occurrence of excessively spiroconvex tests (as the most dominant abnormality type in *Ammonia beccarii*) coincided with the lowest levels of organic compounds. However, those sampling sites with the highest proportions of abnormal tests are situated at the entrance of the Kiel and Flensburg fjords and may be thus influenced by the advection of higher-salinity deep water. Therefore, a high variability of salinity at such settings must be considered (Lutze, 1965, 1974).

3.5.2 Predominance of small test abnormalities

In both fjords, a predominance of small test deformations, such as reduced and overdeveloped chambers, was observed. Myers (1943) reported reduced chamber size in foraminiferal tests and noted that chambers formed during the winter are often smaller than those added during the summer, thus leading to an irregular test shape. These circumstances could be valid for the Kiel Fjord, where samples, containing foraminifera with reduced chambers were taken in winter. However, this argument does not account for the excessive number of small test abnormalities in Flensburg Fjord, where sampling took place in summer.

Nevertheless, small test deformations presumably do not reflect long-term environmental stress, and may only correspond to transient local disturbances in the environment (Geslin and others, 2000). It therefore appears that occasional salinity changes, due to saltwater intrusion from the Belt Sea, have a transient effect on environmental conditions at the entrance of both the Kiel and Flensburg Fjords, where a predominance of small test abnormalities was observed.

3.5.3 Morphological constraints for development of the abnormal tests

The results obtained in this study pose the question whether some test abnormalities are species specific. According to our findings, excessively spiroconvex tests and bulla-like chambers at the umbilicus are typical for *Ammonia beccarii* only. It might well be that, related to the peculiarities of test morphology, there are only limited possibilities for development of test abnormalities. Similar assumptions came from the moving reference model of foraminiferal tests development (Tyszka and others, 2005; Tyszka, 2006). These authors reported that tests develop in a normal way only in a certain morphophase. As applied to *A. beccarii,* by morphophase we mean the trochospiral morphology of its test. Once test development approaches a transitional area, for example, distinct change from trochospiral to planispiral or irregular morphophase, even a slight change in parameters (deviation angle, rotation angle or chamber scaling rate) can lead to a radical change in test morphology (Tyszka, 2006). This morphological feature allows *Ammonia* spp. to develop the species-specific abnormality types, such as bulla-like chambers and excessively spiroconvex tests, which certainly do not occur in species with planispiral tests. The formation of species-specific types of test abnormality is therefore consistent with the concept of a "vacant range" (Tyszka, 2006) of a certain morphophase, which is therefore limited. That means that species with trochospiral morphology have limited possibilities for the development of different abnormality types.

3.5.4 Aberrant tests of *Ammonia beccarii* in Gelting Bay

Shell loss

Buzas-Stephens and Buzas (2005) reported three possibilities for shell loss in living specimens: predation, abrasion and test dissolution. In the western Baltic Sea, abrasion and dissolution were considered to be responsible for shell loss (Grobe and Fütterer, 1981). Walker (1991) observed etched foraminiferal tests after ingestion of foraminifera by periwinkle *Littorina littorea*, which occurs in both fjords (Worm and Lotze, 2006). He noted that rotaliid specimens were still alive after they have been ingested, subjected to the mechanical and chemical digestion and removal from the gut or faecal pellets of the gastropod. This fact explains why all corroded foraminiferal individuals we observed in Gelting Bay were alive (stained). On the other hand, we found the relatively high foraminiferal population densities (170 ind/10cm^3 on average) in the sandy sediments of Gelting Bay, in spite of low food availability (TOC - 0.4%, SiO_2 - 1%, Chl *a* - 11 mkg/g on average) as compared to the inner fjord (TOC - 11%, SiO2 - 6%, Chl a - 108 mkg/g on average). Under similar conditions, Buzas and others (1989) did an experiment on predation and observed very low foraminiferal population densities (1.5 ind/10 cm3 maximum). Therefore if predation played a significant role in the shell loss, it would affect also the foraminiferal abundances though they were quite high in our case. At the same time we may hypothesize the passive ingestion of *Ammonia beccarii* by gastropod *Littorina littorea*, indicating that this foraminiferal species is apparently not exploited as a food source.

Analysis of SEM images revealed that corroded specimens with visible inner organic lining (Pl. 6, Figs. 2-5) most likely result from test dissolution (E. Alve, written communication, 2007). A similar situation was observed at pH ~ 7 in Sandebukta (Oslo Fjord) by Alve and Nagy (1986), who reported dissolved *Ammonia batavus* associated with seasonally enhanced dissolution of calcareous tests between April and June. In the south-western Baltic Sea, similar dissolution processes driven by seasonality were described in the Eckernförde Bay, where foraminiferal tests dissolve in the uppermost millimetres of the sediment (Wefer, 1976). Elastic inner organic linings of foraminiferal tests, folded in places (Pl. 6, Fig. 6), were found in Gelting Bay at water depths of 5-8 m, in agreement with Alve and Nagy (1986), who recorded them within the depth range of 4-11 m.

Since the dissolution phenomenon was unknown to authors at the time of sampling, no measurements of pH values in bottom and pore waters were conducted in Flensburg Fjord, and we can only speculate that a lowering of pH could induce shell loss. In the majority of corroded foraminiferal tests, the last chambers were missing (Pl. 6, Figs. 2, 3). Le Cadre and others (2003) reported loss of the final and thinnest chambers as an initial stage of test dissolution at a pH of seven during culture experiments.

Shell loss in the Flensburg Fjord occurs exclusively in tests of *Ammonia beccarii* – up to 100% of *Ammonia* individuals we found were corroded. As it was shown by experimental studies for *Ammonia tepida* (Stouff and others, 1999b; Le Cadre and others, 2003; Le Cadre and Debenay, 2006), a specimen of this species is able to thicken its inner organic lining under unfavourable conditions and thus regenerate a damaged test. We observed the traces of such regeneration in individuals of *A. beccarii* from the Kiel Fjord (Pl. 4, Fig. 5). At the same time *Ammonia* spp. have different modes of life: epifaunal, infaunal, epiphytic (Debenay and others, 1998; Murray, 2006). Therefore, an infaunal mode of life may be also responsible for shell loss due to changes in pore water pH.

Chapter 3 – Test abnormalities

Another possible reason for shell loss is eutrophication, which lowers oxygen levels and makes pore waters anoxic and sulphidic (Jorissen, 1999), affecting simultaneously dissolution processes and bioavailability of metals. However, we did not observe the high levels of organic matter in Gelting Bay, as compared to inner Flensburg Fjord (see above). In addition, sandy sediments in this area do not provide the accumulation of heavy metals and organic matter. Thus, we may exclude eutrophication from reasons, inducing shell loss in Gelting Bay.

Exon (1971) reported that the eastern Gelting Bay is an area exposed to intensive bottom currents of 30 cm/s that cause active westward long-shore sediment drift. Transport of such strong currents can damage living benthic foraminiferal tests, causing abnormal test shapes to form during the regeneration of damaged chambers (Geslin and others, 2002). Nevertheless, some corroded shells of the *Ammonia beccarii* had only interlocular walls. These so-called "star-shaped" tests (Pl. 6, Fig. 5) have been reported after heavy dissolution (Buzas-Stephens and Buzas, 2005; Le Cadre and others, 2003). This peculiar test shape is easily distinguishable from the shell breaks caused by abrasion, which also destroys interlocular walls. We therefore conclude that dissolution is the main process responsible for the shell loss seen in Flensburg Fjord.

Multiple tests

A sample, taken off the Gelting Noor showed a distinct abnormality type in *Ammonia beccarii* resembling a fusion of two specimens (Pl. 7, Figs 5, 6, 8). A smaller individual displayed extremely thin walls and disruptions in the coiling plane. The bigger specimen showed thick but corroded walls, creating an illusion sometimes that a new foraminifer comes out of the older, corroded test. Sharifi and others (1991) reported similar double specimens from Southampton Water (UK), which also differed in size. These authors noted that double specimens of the same size are very rare.

Culture experiments under hypersaline conditions (Stouff and others, 1999a, b) revealed the following reasons for double or multiple tests. Firstly, an anomaly in the development of a single juvenile may cause the formation of double tests. In this case, two protuberances can form on the uncalcified proloculus of a young specimen, giving a way to the formation of two second chambers with subsequent development of whorls from each of the second chambers. Secondly, multiple tests result from an early fusion of two young specimens which both continue their development. Finally, attachment of the juvenile to a parental test may occur after schizogony. After fusion, the attached juvenile continues its development creating thus a double test.

On the other hand, Wennrich and others (2007) suggested the formation of double tests is caused by hyposalinity. The area off the Gelting Noor faces the saltwater inflows from the Danish

Chapter 3 – Test abnormalities

Straits (Exon, 1971). Foraminifera dwelling in the Gelting Bay may undergo drastic and sudden changes in salinity. If the timing of saltwater intrusion coincides with periods of foraminiferal reproduction, abrupt salinity changes may prevent dispersal and facilitate the early fusion of juveniles after schizogony (Stouff and others, 1999a, b), leading to the formation of abnormal tests.

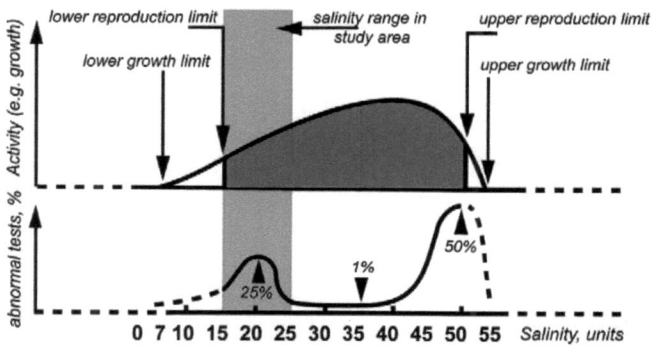

Figure 3.5: Conceptual model illustrating the salinity tolerance curve of *Ammonia* spp. and salinity range for development of abnormal tests.

3.5.5 Abnormalities as a result of abrupt salinity changes

Ammonia beccarii was described by Wefer (1976) as a euryhaline species that ranges from hypo- to hypersalinities. This species provides the majority of specimens with abnormal tests at the sites affected by salt-water intrusion in both fjords. In addition, the spatial distribution of all groups of abnormalities in both fjords points at the higher frequencies in areas with salinity variations, and namely outer parts, Gelting Bay and Schwentine mouth. Therefore, we suggest that critical thresholds of salinity induce the development of foraminiferal test abnormalities. Based on our findings and previous studies (see capture to Fig. 3.5), we propose a conceptual model illustrating the relationship between the normal development of representatives of the genus *Ammonia* and formation of abnormal tests. (Fig. 3.5).

Our results confirm the hypothesis that hyposalinity may be responsible for development of abnormal tests. Even if the observed salinity range is within the tolerance of a certain species, salinity levels close to the lower reproduction limit (Fig. 3.5) may induce the formation of aberrant tests, owing to the enhanced vulnerability of the juvenile specimens. The salinity needs only to rise or fall suddenly, as happens often in the western Baltic Sea, to facilitate the formation of abnormal tests.

The overwhelming majority of specimens found in current study was megalospheric and represented therefore the offsprings produced asexually (Alve and Goldstein, 2003). Megalospheric tests prevail in most natural populations (Leutenegger, 1977; Lehmann and others, 2006) and often indicate the asexual reproduction mode, as more appropriate under the unfavourable conditions (Nigam and Caron, 2000; Coccioni, 2000). Vice versa, the culture experiment with *Rosalina leei* showed that an increase in temperature resulted in an increase in tendency for sexual reproduction (Nigam and Caron, 2000). In our case, a predominance of offsprings produced asexually may reflect the inflows of more cold and salt-rich bottom water from the Belt Sea. Elevated frequencies of abnormalities may thus indicate a population surviving highly variable environmental conditions close to the species' reproduction limit.

3.6 CONCLUSIONS

We recognized 18 types of foraminiferal test abnormalities in the Kiel and Flensburg Fjords of the Kiel Bay. According to morphological criteria, the types were classified into five groups: chamber, apertural, umbilical, coiling and test abnormalities. In both fjords, a predominance of small test deformities such as reduced and overdeveloped chambers, was observed. These small test deformations are here related to short-term environmental changes. In particular, intrusions of salt-rich bottom waters from the Belt Sea are the cause of the highest proportions of abnormal foraminiferal tests in the outer parts of both fjords. In the inner fjords, elevated levels of heavy metals apparently lead to high percentages of abnormal tests. Our data show different relationships of abnormal tests with heavy metals in both fjords due to different hydrographical conditions.

In both fjords, test abnormalities are over-represented in *Ammonia beccarii* and under-represented in *Elphidium excavatum* subspecies compared to their average proportions in the living assemblages. A bulla-like chamber, covering the umbilicus, and excessively spiroconvex tests were seen only in *A. beccarii*. These species-specific abnormalities were explained by the limited possibilities for the abnormal development due to trochospiral morphology of *Ammonia* spp.

Tests of *Ammonia beccarii* found in Flensburg Fjord showed distinct irregularities reflecting dissolution and development of double tests due to special environmental conditions in the Gelting Bay, where changes in salinity and enhanced sediment redeposition prevail. Our study confirms the hypothesis that within a certain area natural instability might be more important than anthropogenic influence for development of abnormal tests. Even if environmental variability is well within the range of tolerance of a given species, such environmental changes may cause abnormalities if coincident with a period of high sensitivity of the organisms (e.g., during reproduction). Thus, using abnormal foraminiferal tests as an indicator of environmental pollution must be done with care, especially in settings exhibiting unstable environmental conditions.

ACKNOWLEDGEMENTS

The authors are grateful to crew of RV Polarfuchs Holger Meyer and Helmut Schramm for help with sampling. We are indebted to Anna Nikulina, who provided the data on sediment geochemistry and helped with foraminiferal analysis of samples. We acknowledge Udo Laurer (IFM-GEOMAR), Ute Schuldt, Claudia Ehlert (Institute of Geosciences, Kiel) and Irina Gembitskaya (St. Petersburg State Mining Institute) for technical assistance. We also wish to thank Rodolfo Coccioni (University of Urbino), anonymous reviewer, Charlotte Brunner (University of Southern Mississippi) and Pamela Hallock (University of South Florida) for constructive comments, valuable suggestions and reviewing of this manuscript. Our sincere thanks go to Elisabeth Alve (University of Oslo) for her valuable advises during the early stage of this study. Finally, we want to express our gratitude to Brian Haley (IFM-GEOMAR) for linguistic corrections of earlier version of this manuscript. This study was funded by Leibniz Award DFG DU 129-33 and research grant of Otto-Schmidt Laboratory (AARI, St. Petersburg).

Outlook

Preparing the first article on foraminiferal response to the recent environmental changes in Kiel Fjord, we encountered the problem of missing background or reference values for trace metal and organic compound concentrations. The Baltic Sea has a very long history of pollution and therefore it is difficult to find out baseline values for natural conditions. For this purpose, another approach is appropriate, which is widely used in paleoceanography: we investigated a high-resolution sediment record, which goes back to times with minimum anthropogenic impact.

In order to establish such reference conditions, a short core was taken in the outer part of Kiel Fjord (double cores PF17-38). The outer part of the fjord was chosen, because previous studies showed that the inner fjord bottom was exposed to extended dredging during construction of harbours and piers in 1960-70s and constant sediment redeposition due to high ship traffic (Schwarzer & Themann, 2003; LANU, unpubl. data). The core from the outer Kiel Fjord showed a pronounced stratification and was therefore considered as undisturbed and suitable for further analysis. This sediment record will allow going back to the 1860s, when steam ships were invented. As a proxy for the "steam-shipping era", the occurrence of ship clinker in the sediments was established.

The first 10 cm of a core PF17-38 were sliced downward in 0,5 centimetre-interval for high-resolution foraminiferal analysis. The rest part (from 10 to 40 cm) was sliced in 2 cm-intervals. Samples were washed out through 63-μm sieve at first with solution of ammonia and distilled water and the first flush of <63-μm size fraction was collected into the PVC vials for geochemical analyses. The samples were washed further with tap water for benthic foraminifera, dried at 60°C, splitted and picked.

A duplicate of PF17-38 was subsampled until 10 cm depth for chronological analysis. The upper 5 centimetres were sliced in 0.5 cm-intervals. In the following 5-10 cm interval, every centimetre was taken. Sediment subsamples were freeze-dried and powdered. Dating of the core material was carried out by non-destructive counting of the ^{137}Cs and ^{210}Pb γ-ray activity in Laboratory for Radioisotopes (LARI, Göttingen). Anna Nikulina, who is currently processing the data on ^{137}Cs and ^{210}Pb activity, provided a tentative age model and geochemical data for this core.

First results of geochemical analysis showed an increase of organic carbon and total nitrogen in the 1960s, followed by a decrease in the late 1970s, when sewage treatment plant Bülk was built and canalisation outlet at the outer fjord was set into operation (Kallmeyer, 1997).

Preliminary results of foraminiferal analysis showed four distinct peaks of arenaceous *Ammotium cassis* at depths 2.3, 4.3, 6.8 and 8.8 cm, which correspond approximately to the 1930s, 1950s, 1970s and 1990s. These peaks were observed within depth interval 0-10 cm, which

Outlook

was sampled in high resolution. After 10 centimetres the distribution of *A. cassis* showed a less clear pattern (Fig. 4.1). Nonetheless, this arenaceous species was present in Kiel Fjord before the 1930s. These results are contradictory with conclusions of Olsson (1976), who suggested that *A. cassis* colonized the southern part of the Baltic Proper not earlier than in the 1960s. Such speculation was based on previous studies from Gullmar Fjord (Höglund, 1947), Oslo Fjord (Christiansen, 1958) and SW Baltic (Rhumbler, 1935). On the basis of one core, it is difficult to make conclusions why *A. cassis* was abundant in Kiel Fjord before the 1930s. All the more so, as the lower part of the core was sampled in lower resolution (every two cm), as compared to the upper part, any considerations of fluctuations are speculative.

According to Schönfeld and Numberger (2007a), a decline of *A. cassis* in Eckernförde Bight and the outer Kiel Fjord is explained by low salinity events in the late 1930s and early 1990s, which confine the period when this species was abundant in the Kiel Bight. We compared our peaks of *A. cassis* to the continuous records of salinity for the Kattegat water at the Koljö Fjord and Gotland Basin and to the occurrence of major Baltic inflows (MBIs) (Filipsson & Nordberg, 2004; Matthäus, 2006).

The results (Fig. 4.1) showed that, indeed, the periods of lower salinity and lack of MBIs coincide with absence of *A. cassis* in the outer Kiel Fjord. An exception represents the last frequent occurrence of *A. cassis* around 1990s, which corresponds to a higher salinity event in the Koljö Fjord, but to the absence of MBIs, simultaneously. Taking into account a preliminary assessed dating error of several years, we may speculate that this peak of *A. cassis*, in time coincides with a 1996-peak, reported in outer Kiel Fjord by Schönfeld and Numberger (2007a). In this case, it followed, apparently, a salt-water inflow, which occurred in 1993.

Over the past two decades the frequency of major salt-water inflows has decreased significantly (Matthäus, 2006; Meier et al., 2004; Feistel et al., 2007) and recently there were only three major Baltic inflows - in 1993, 1997 and 2003. Apparently, they were insufficient for an establishment of a stable halocline. This, however, is needed for nutrition of *A. cassis*, and therefore impeded the recolonisation of this species in the Kiel Bight. As for detailed record of this arenaceous species before the 1930s, the high-resolution analysis of additional cores from the Kiel Bight is needed.

Outlook

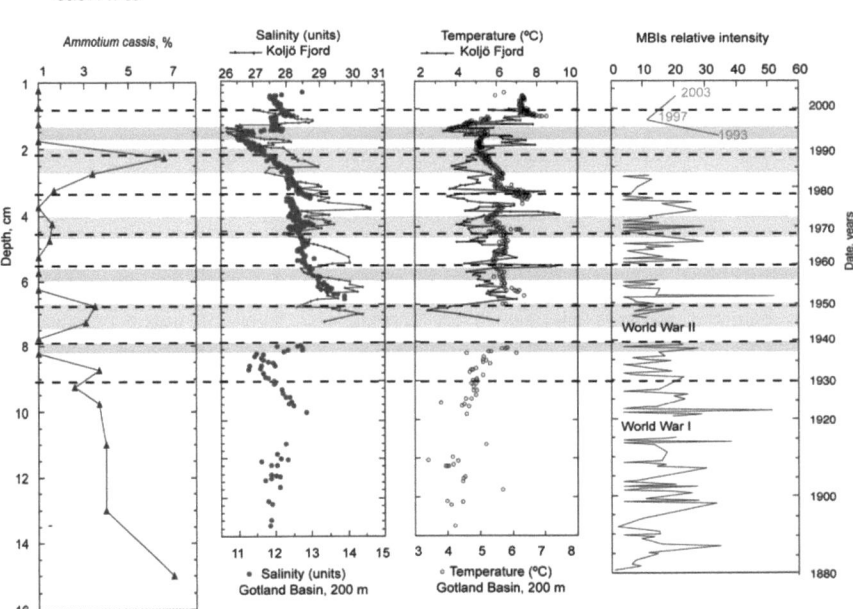

Figure 4.1: Occurrence of arenaceous *Ammotium cassis* down core PF17-38, taken in outer Kiel Fjord. Salinity and temperature records are modified after Filipsson & Nordberg (2004) and Matthäus (2006). Both they are given for the Koljö Fjord (Sweden) and the Gotland Basin. Relative intensity record of the major Baltic inflows (MBIs) is modified after Matthäus (2006). Green shaded areas emphasize the peaks in abundances of *A. cassis*, whereas grey ones show periods when this species was absent in outer Kiel Fjord.

General conclusions

Baltic Sea is one of the largest brackish marginal seas in the world. Its ecosystem is highly vulnerable, due to a high environmental variability. This is reflected in the occasional ventilation of deep-water by high saline water from the Kattegat through the narrow and shallow Danish Straits. During the past two decades the frequency of major inflows has decreased significantly, which has promoted extended periods with decreasing bottom water oxygenation. At the same time, a growing economy led to significant anthropogenic pollution in this area, which caused eutrophication, oxygen depletion, and elevated levels of trace metals in the bottom sediments. These environmental changes are seen today at all trophic levels of marine organisms and affected even one of the smallest inhabitants of the Baltic Sea: benthic foraminifera.

Within the framework of this thesis, 109 surface sediment samples were analysed for living (Rose Bengal stained) benthic foraminifera. Ten taxa (eight calcareous and two arenaceous) were found. Foraminifera were evaluated as proxies of recent environmental change both of natural and anthropogenic origin in two fjords of the Kiel Bight (SW Baltic Sea). Kiel Fjord, moderately polluted by trace metals, was chosen as study area for tracing the anthropogenic influence on benthic foraminifera. Foraminiferal assemblages of Flensburg Fjord were surveyed for a first time and they reflected a high natural variability, which is linked to occasional salt-water inflows from the North Sea. A validity to use the abnormal foraminiferal tests as indicators of anthropogenic pollution is discussed. My approach allowed developing the conceptual model showing the relationship between salinity tolerance of certain foraminiferal species and the development of test abnormalities.

The main conclusions from this thesis are as follows:

Foraminiferal response to environmental change in Kiel Fjord.

Analysis of foraminiferal population density showed a patchy distribution and a response to food availability, which was depicted by SiO_2 and Chl *a* in the sediments. Markedly low levels of food particles in Friedrichsort Sound established quite unfavourable conditions for many benthic foraminiferal species. A strong increase of population density since the 1960s remained enigmatic. It cannot be attributed to an increase in organic matter supply and a slight reduction of pollution. Furthermore, significant changes in foraminiferal species composition in 2005-2006 as compared to the 1960s were recorded. The highly stress tolerant species *A. beccarii* invaded Kiel Fjord and replaced *Ammotium cassis*. Unfavourable salinity conditions in the Kiel Bight and absence of a deep halocline in Kiel Fjord might have caused the disappearance of *A. cassis*. We suppose that *A. beccarii* is highly opportunistic and capable to tolerate elevated levels of nutrients and trace metals. On the other hand, *E. albiumbilicatum* apparently was able to withstand the higher water turbulences and therefore inhabited the transitional area of Friedrichsort Sound. During the winter season, linkages between test abnormalities and trace metal concentrations were not obvious. However, during springtime we observed an increase in abundance of abnormal tests, which was correlated to high trace metal levels. This mirrored the reproduction of benthic foraminifera during spring bloom and showed that juveniles were especially sensitive to environmental stress. The dominance of *Ammonia beccarii* was not affected by the increased abundance of abnormal juvenile specimens.

Recent benthic foraminifera in Flensburg Fjord

Five foraminiferal biofacies were distinguished in Flensburg Fjord. Their distribution appeared to be controlled mainly by food availability, when oxygen was not a limiting factor for foraminifera. The inner Flensburg Fjord (Biofacies 1) was dominated by *Elphidium incertum* dwelling within muddy sediments rich in food particles. Biofacies 2 comprised the "*E. incertum – E. excavatum*" group, which was found in muds and sandy-muds of the fjord loop around Holnis Peninsula and in the outer fjord at less food availability. The Gelting Bay depicts a distinctively different area with shallow-water, brackish and sandy habitat poor in food particles. This area is inhabited by the assemblage of Biofacies 3, which is dominated by *Ammonia beccarii* and *Elphidium albiumbilicatum*. *A. beccarii*, *E. excavatum* subspecies and *E. incertum* were dominant in the central Flensburg Fjord and near-shore zones of loop (Biofacies 4) with sandy muds relatively poor in food particles. *Elphidium excavatum* subspecies inhabited the innermost part of the fjord (Biofacies 5) with the fine-grained muddy sediments rich in food.

Response to environmental change in Flensburg Fjord

It was suggested that the frequent occurrence of the infaunal *Elphidium incertum* in the uppermost sediment layer might reflect seasonal anoxic conditions in the inner Flensburg Fjord, that have preceded the sampling period. On the other hand, an A/E index exceeding 80% and enhanced porosity of *Ammonia beccarii* tests apparently mirror an oxygen deficiency in the sheltered part of Gelting Bay.

A comparison with previous studies from late 1940s and 1970s revealed substantial changes in species composition in the outer Flensburg Fjord with a decline of arenaceous *Ammotium cassis*, flourishing of calcareous *Ammonia beccarii* in the Gelting Bay and dominance of *Elphidium incertum* in the inner fjord. These changes are similar to those, reported recently from other fjords of the SW Baltic Sea (Eckernförde Bay and Kiel Fjord). They are most likely associated with a generally decreased intensity and frequency of major Baltic inflows since the 1960s.

Conclusions

Foraminiferal test abnormalities

Eighteen modes of abnormal foraminiferal tests were detected in the Kiel and Flensburg fjords of the Kiel Bight. According to morphological criteria, all modes were classified into 5 groups: chamber-, aperture-, umbilicum-, coiling- and test abnormalities. In both fjords, a predominance of so called "small test deformations" such as reduced chamber size and overdeveloped chambers, was observed. These "small test deformations" indicate short-term environmental changes. In particular, inflows of salt-rich bottom waters from the North Sea are the background reasons for the highest proportions of abnormal foraminiferal tests in outer parts of both fjords. In the inner fjords, elevated levels of heavy metals may lead to high percentages of abnormal tests. Our data indicate a non-linear relationship of abnormal tests with heavy metals and correlations of certain abnormality modes at elevated, but not extreme, levels of pollution.

In both fjords test abnormalities are over-represented in *Ammonia beccarii* and under-represented in *Elphidium excavatum* subspecies compared to their average proportions in the living assemblages. Bulla-like chamber, covering the umbilicum, and spiroconvex tests with a distinctly high spiral side were seen only in *A. beccarii*. These species-specific abnormalities were explained by the limited "vacant range" during the "evolution" of the trochospiral morphophase, following Tyszka (2006).

Tests of *A. beccarii* found in Flensburg Fjord showed distinct irregularities reflecting dissolution and development of double tests due to special environmental conditions in Gelting Bay, where changes in salinity and enhanced sediment redeposition prevail. On this basis, we developed a conceptual model showing the relationships between salinity tolerance of *Ammonia* species and development of abnormal tests. This model demonstrates that not only hypersaline conditions are responsible for the higher frequencies of abnormal tests. Indeed, a hyposalinity also plays an important role in the development of foraminiferal tests abnormalities.

This study confirms the idea that within a certain area, natural instability might be more important than anthropogenic influence for development of foraminiferal test abnormalities. Even if environmental variability is well within the range of tolerance of a given species, such environmental changes may cause abnormalities if they coincide with a period of high sensitivity of the organisms (e.g. during reproduction). Thus, using abnormal foraminiferal tests as an indicator of environmental pollution must be done with care, especially in settings exhibiting unstable environmental conditions.

References

Abu-Hilal A.H., and Badran M.M. (1990) Effect of pollution sources on metal concentration in sediment cores from the Gulf of Aqaba (Red Sea). *Marine Pollution Bulletin* **21** (4), 190-197.

Almogi-Labin A., Perelis-Grossovicz L., and Raab M. (1992) Living *Ammonia* from a hypersaline inland pool, Dead Sea area, Israel. *Journal of Foraminiferal Research* **22**, 257-266.

Alloway, B.J. (2005) Bioavailability of elements in soil *in* Sellinus, O., Alloway, B.J., Centeno, J.A., Finkelman, R.B., Fuge, R., Lindh, U., and Smedley, P. (eds.), Medical geology: Impacts of the natural environment on public health: Elsevier, San Diego, p. 347–372.

Altenbach A.V. (1985) Die Biomasse der benthischen Foraminiferen. Auswertungen von "Meteor"-Expeditionen im östlichen Nordatlantik. PhD. Thesis., CAU-Kiel, 167 p.

Altenbach A.V. (1992) Short-term processes and patterns in foraminiferal response to organic flux rates. *Marine Micropaleontology* **19**, 119-129.

Alve E. (1991) Benthic foraminifera in sediment cores reflecting heavy metal pollution in Soerfjord, Western Norway. *Journal of Foraminiferal Research* **21**, 1-19.

Alve E. (1999) Benthic foraminiferal responses to estuarine pollution: a review. *Journal of Foraminiferal Research* **25**, 190-203.

Alve E., and Nagy J. (1986) Estuarine foraminiferal distribution in Sandebukta, a branch of the Oslo Fjord. *Journal of Foraminiferal Research* **16**, 261-284.

Alve E. and Murray J.W (1999) Marginal marine environments of the Skagerrak and Kattegat: a baseline study of living (stained) benthic foraminifera. *Palaeogeography Palaeoclimatology Palaeoecology* **146**, 171-193.

Alve E. and Olsgardt F. (1999) Benthic foraminiferal colonisation in experiments with copper-contaminated sediments. *Journal of Foraminiferal Research* **29**, 186-195.

Alve E., and Goldstein, S. (2003) Propagule transport as a key method of dispersal in benthic foraminifera (Protista): Limnology and Oceanography, **48**, .2163-2170.

Angel D.L., Verhese S., Lee J.J., Saleh A.M., Zuber D., Lindell D., and Symons A. (2004) Impact of a net cage fish farm on the distribution of benthic foraminifera in the northern Gulf of Eilat (Aqaba, Red Sea). *Journal of Foraminiferal Research*, **30**, 54-65.

References

Ankley, G.T., Thomas, N.A., Di Toro, D.M., Hansen, D.J., Mahony, J. D., Berry, W.J., Swartz, R.C., Hoke, R.A., Garrison, A.W., Allen, H.E., Zarba, C.S. (1994) Assessing potential bioavailability of metals in sediments: a proposed approach: Environmental management, **18**, 331-337.

Aßmann C., Golosnoy V., and Hogrefe J. (2007) Formelsammlung zur Methodenlehre der Statistik II, β - version, Institut für Statistik und Ökonometrie der CAU-Kiel, Germany, 63 p.

Balzer W., Erlenkeuser H., Hartmann M., Müller P.J., and Pollehne F. (1987) Diagenesis and exchange processes at the benthic boundary. In *Seawater—Sediment Interactions in Coastal Waters* (eds. J. Rumohr, E. Walger, and B. Zeitzschel), pp. 111–161. Springer-Verlag, Berlin.

Barry, P.S.I. (1975) Lead levels in blood: Letters to Nature, **258**, 775

Bellinger E.G., and Benham B.R. (1978) The levels of metals in dock-yard sediments with particular reference to the contributions from ship-bottom paints. *Environmental Pollution* **15** (1), 71-81.

Bergin F., Kucuksezgin F., Uluturhan E., Barut I.F., Meric E., Avsar N., and Nazik A. (2006) The response of benthic foraminifera and ostracoda to heavy metal pollution in Gulf of Izmir (Eastern Aegean Sea). *Estuarine Coastal and Shelf Science* **66**, 368-386.

Bernardez P., Frances G. and Prego R. (2006) Benthic-pelagic coupling and postdepositional processes as revealed by the distribution of opal in sediments: The case of the Ria de Vigo (NW Iberian Peninsula). *Estuarine Coastal and Shelf Science* **68**, 271-281.

Bernhard, J.M. (1988). Postmortem vital staining in benthic foraminifera: duration and importance in population and distributional studies. *Journal of Foraminiferal Research*, **18**: 143-146.

Bernhard, J.M., Ostermann, D.R., Williams, D.S. & Blanks, J.K (2006). Comparison of two methods to indetify live benthic foraminifera: a test between Rose Bengal and Cell Tracker Green with implications for stable isotope paleoreconstructions. *Paleoceanography*, **21**: PA4219, doi:10.1029/2006PA001296.

Boltovskoy E., and Wright R. (1976) Recent foraminifera. (ed. W. Junk), The Hague, 515 p.

Boltovskoy E., Scott D.B. and Medioli F.S. (1991) Morphological variations of benthic foraminiferal tests in response to changes in ecological parameters: a review. *Journal of Paleontology* **65**, 175-185.

Bradshaw J. (1957) Laboratory studies on the rate of growth of the foraminifer "Streblus beccarri (Linne) var. tepida (Cushman)". *Journal of Paleontology* **31**, 1138-1147.

References

Bradshaw J. (1961) Laboratory experiments on the ecology of foraminifera. *Contributions from the Cushman Foundation for Foraminiferal Research* **7 (3)**, 87-106.

Brunner C.A., Beall J.M., Bentley S.J., and Furukawa, Y. (2006) Hypoxia hotspots in the Mississippi Bight. *Journal of Foraminiferal Research* **36**, 95-107.

Brügmann L. (1996) Quellen und regionale Verteilung von Schwermetallen im Wasser und Sediment. In *Warnsignale aus der Ostsee* (eds. J.L. Lozan, R. Lampe, W. Mätthaus, E. Rachor., H. Rumohr, H. von Westernhagen), pp. 74-79. Parey Buchverlag, Berlin.

Burone L., Venturini N., Sprechmann P., Valente P., and Muniz P. (2006) Foraminiferal responses to polluted sediments in the Montevideo coastal zone, Uruguay. *Marine Pollution Bulletin* **52**, 61-73.

Buzas, M.A., Collins, L.S. Richardson, S.L. and Severin, K.P. (1989) Experiments on predation, substrate preference, and colonization of benthic foraminifera at the shelfbreak off the Ft. Pierce Inlet, Florida: Journal of Foraminiferal Research, **19**, 146-152.

Buzas-Stephens P. and Buzas M.A. (2005) Population dynamics and dissolution of Foraminifera in Nuices Bay, Texas. *Journal of Foraminiferal Research* **35**, 248-258.

Chandler G.T. (1989) Foraminifera may structure meiobenthic communities. *Oecologia*, **81**, 354-360.

Chester, R. and Aston, R. (1976) The geochemistry of deep-sea sediments *in* Riley, J.P. and Chester, R., (eds,) Chemical Oceanography v. 6: Academic Press, London, p. 281–390.

Clark E.A., Sterritt R.M., and Lester J.N. (1988) The fate of tributyltin in the aquatic environment: a look at the data. *Environmental Science and Technology* **22** (6), 600-604.

Christiansen B. (1958) The foraminifer fauna in the Dröbak Sound in the Oslo Fjord (Norway), *Nytt. Mag. Zool* **6**, 5-91.

Coccioni, R. (2000) Benthic foraminifera as bioindicators of heavy metal pollution *in*: Martin, R.M. (ed.) Environmental micropaleontology: The applications of microfossils to environmental geology, Springer, p. 71-103.

Corliss, B.H. & Emerson, S. (1990). Distribution of Rose Bengal stained deep-sea benthic foraminifera from the Nova Scotian continental margin and Gulf of Maine. *Deep-Sea Research*, **37**: 381-400.

Debenay, J.-P., Beneteau, E., Zhang, J., Stouff, V., Geslin, E., Redois, F., and Fernandez-

References

Gonzalez, M. (1998) *Ammonia beccarii* and *Ammonia tepida* (Foraminifera): morphofunctional arguments for their distinction: Marine Micropaleontology, **34**, 235-244.

Debenay J.-P., Tsakiridis E., Soulard R., Grossel H. (2001) Factors determining the distribution of foraminiferal assemblages in Port Joinville Harbour (Ile d'Yeu, France): the influence of pollution. *Marine Micropaleontology* **43**, 75-118.

De Nooijer L. (2007) Shallow-water benthic foraminifera as proxy for natural versus human-induced environmental change 2007. Geologica Ultraiectina, *Mededelingen van de Faculteit Geowetenschappen Universiteit Utrecht*, **272**, 152 pp.

De Nooijer, L., Duijnstee, I.A.P. & van der Zwaan, G.J. (2006). A novel application of MTT reduction: a viability assay for temperate shallow-water benthic foraminifera. *Journal of Foraminiferal Research*, **36**: 195-200.

De Nooijer, L. J., Duijnstee, I.A.P., Bergman, M.J.N. & van der Zwaan, G.J. (2008). The ecology of benthic foraminifera across the Frisian Front, southern North Sea. *Estuarine, coastal and shelf science*, **78**: 715-726

Di Leonardo R., Bellanca A., Capotondi L., Cundy A., and Neri R. (2007) Possible impacts of Hg and PAH contamination on benthic foraminiferal assemblages: an example from the Sicilian coast, central Mediterranean. *Science of the Total Environment* **388**, 168-183.

DDTFF (1992) Deutsch-Dänische Technikergruppe für die Flensburger Förde: Jahresbericht 1992, Apenrade, 42 p.

Ellison R., Broome R., and Ogilvie R. (1986) Foraminiferal response to trace metal contamination in the Patapsco River and Baltimore Harbour, Maryland. *Marine Pollution Bulletin* **17**, 419-423, 1986.

Erlenkeuser H., Suess E., Willkomm H. (1974) Industrialization affects heavy metal and carbon isotope concentrations in recent Baltic Sea sediments, *Geochimica et Cosmochimica Acta* **38**, 823-842.

Ernst S.R., Morvan J., Geslin E., Le Bihan A. and Jorissen F.J. (2006) Benthic foraminiferal response to experimentally induced Erika oil pollution. *Marine Micropaleontology* **61**, 76-93.

Exon N. (1971) Holocene sedimentation in and near the outer Flensburg Fjord (westernmost Baltic Sea), PhD thesis, CAU Kiel, 102 p.

References

Exon N. (1972) Sedimentation in the outer Flensburg Fjord area (Baltic Sea) since the last Glaciation. *Meyniana*, **22**, 5-6.

Fatela F., and Taborda R. (2002) Confidence limits of species proportions in microfossil assemblages. *Marine Micropaleontology*, **45**, 169-174.

Feistel R., Nausch G., and Hagen E., 2007. Water exchange between the Baltic Sea and the North Sea, and conditions in the deep basins. HELCOM Indicator Fact Sheets 2007. Available online at: http://www.helcom.fi/environment2/ifs/en_GB/cover/

Fennel W. (1996) Wasserhaushalt und Strömung, In *Meereskunde der Ostsee* (ed. G. Rheinheimer), pp. 56-67. Springerverlag, Berlin.

Filipsson H. and Nordberg K. (2004) A 200-year environmental record of a low-oxygen fjord, Sweden, elucidated by benthic foraminifera, sediment characteristics and hydrographic data. *Journal of Foraminiferal Research* **34**, 277-293.

Frenzel P., Tech T., and Bartholdy J. (2005) Checklist and annotated bibliography of Recent Foraminiferida from the German Baltic Sea coast. In *Methods and applications in micropaleontology* (ed. J. Tyszka.). *Studia Geologica Polonica* **124**, 67-86.

Frontalini F., and Coccioni R. (2007) Benthic foraminifera for heavy metal pollution monitoring: a case study from the central Adriatic Sea coast of Italy. *Estuarine Coastal and Schelf Science* **76** (2), 404-417, doi:10.1016/j.ecss.2007.07.024

Förstner, U. (1993) Metal speciation – general concept and application: International Journal of Environmental Analytical Chemistry, **51**, 5-23.

Garbe-Schönberg C.D. (1993) Simultaneous determination of 37 trace elements in 28 international rock standards by ICP-MS. *Geostandard Newsletter* **17**, 81-93.

Gemeinsames Komitee Flensburger Förde (1974) Untersuchungen der Flensburger Förde: Wasseraustausch, Teilbericht 7: Selbstverlag, Amtshuset, Aabenraa, 129 p.

Gerlach S. (1984) Oxygen depletion 1980 - 1983 in coastal waters of the Federal Republic of Germany, *Berichte aus dem Institut fur Meereskunde an der CAU-Kiel* **130**, 97 pp.

Gerlach, S. (1990) Nitrogen, phosphorus, plankton and oxygen deficiency in the German Bight and the Kiel Bay. *Kieler Meeresforschungen* **7**, 341 pp.

References

Gerlach S. (1996) Ökologische Veränderungen in der Kieler Bucht. In *Warnsignale aus der Ostsee* (eds. J.L. Lozan, R. Lampe, W. Matthäus, E. Rachor., H. Rumohr, H. von Westernhagen), pp. 259-292. Parey Buchverlag, Berlin.

Geslin, E., Debenay, J.-P., and Lesourd, M. (1998) Abnormal wall textures and test deformation in Ammonia (hyaline foraminifer): *Journal of Foraminiferal Research*, **28**, 148-156.

Geslin E., Stouff V., Debenay J.-P., and Lesourd M. (2000) Environmental variation and foraminiferal test abnormalities. In *Environmental micropaleontology. The application of microfossils to environmental geology* (ed. R. Martin), pp. 191-215. Kluwer Academic Publishers/Plenum Publishers, New York.

Geslin E., Debenay J.-P., Duleba W., and Bonetti C. (2002) Morphological abnormalities of foraminiferal tests in Brazilian environments: comparison between polluted and non-polluted areas. *Marine Micropaleontology* **45**, 151-168.

GKFF (Gemeinsames Komitee Flensburger Förde) (1973). *Untersuchungen der Flensburger Förde: Wasseraustausch*, Teilbericht 7: Selbstverlag, Amtshuset, Aabenraa, 129 pp.

Graf G., Bengtsson W., Diesner U., Sehulz R. and Theede H. (1982) Benthic response to sedimentation of a spring phytoplankton bloom: process and budget. *Marine Biology* **67**, 201-208.

Greiser N. and Faubel A. (1998) Biotic factors. In *Introduction to the study of meiofauna*, (eds R.P. Higgins and H. Thiel), pp. 79-114. Smithsonian Institution Press, London.

Grobe H. and Fütterer D. (1981) Zur Fragmentierung benthischer Foraminiferen in der Kieler Bucht (Westliche Ostsee). *Meyniana*, **33**, 85-96.

Gustafsson M. and Nordberg K. (1999) Benthic foraminifera and their response to hydrography, periodic hypoxic conditions and primary production in the Koljö fjord on the Swedish west coast. *Journal of Sea Research*, **41**, 163-178.

Gustafsson M. and Nordberg K. (2001) Living (stained) benthic foraminiferal response to primary production and hydrography in the deepest part of the Gullmar Fjord, Swedish west coast, with comparison to Höglund's 1927 material, *Journal of Foraminiferal Research* **31**, 2-11.

Gustafsson, M., Dahllöf, I., Blanck, H., Hall, P., Molander, S., and Nordberg, K. (2000) Tri-n-butyltin (TBT) pollution in an experimental mesocosm: Marine Polllution Bulletin, **40**, 1072-1075.

References

Haarich M., Pohl C., Leipe T., Grünwald K., Bachor A., Weber M., Petenati T., Schröter-Kermani C., Jansen W. und Bladt, A. (2003) Anorganische Schadstoffe. In *Meeresumwelt 1999-2002. Ostsee.* Bund-Länder-Messprogramm für die Meeresumwelt von Nord- und Ostsee, Kiel, pp. 167-194.

Hallock, P. (2000). Larger foraminifera as indicators of coral-reef vitality, *In:* Martin, R.E. (Ed.), Environmental micropaleontology: The application of microfossils to environmental geology. *Topics in biogeology,* **15**: 121-150.

Hayward B.W., Holzmann M., Grenfell H.R., Pawlowski J., and Triggs C.M. (2004) Morphological distinction of molecular types in *Ammonia* – towards a taxonomic revision of the world's most commonly misidentified foraminifera, *Marine Micropaleontology* **50**, 237-271.

Heeger T. (1990) Elektronenmikroskopische Untersuchungen zur Ernährungbiologie benthischer Foraminiferen. *Berichte Sonderforschungbereich 313*, CAU-Kiel, **21**, 1-139.

HELCOM (1993) First Assessment of the State of the Coastal Waters of the Baltic Sea, *Baltic Sea Environmental Proceedings* **54**, Helsinki, 166 pp.

Helland A., and Bakke T. (2002) Transport and sedimentation of Cu in microtidal estuary, SE Norway. *Marine Pollution Bulletin* **44**, 149-155.

Hermelin J. O. (1987) Distribution of Holocene benthic foraminifera in the Baltic Sea. *Journal of Foraminiferal Research* **17**, 62–73.

Höglund H. (1947) Foraminifera in the Gullmar Fjord and the Skagerrak. *Zool. Bidr. Upps.* **26**, 1-328.

IMO (2005) Anti-Fouling Systems. International Convention on the Control of Harmful Anti-Fouling Systems on Ships, 2005 edition. London. Available online at: http://www.imo.org/Conventions/mainframe.asp?topic_id=529

Kallmeyer T. (1997) Probleme der Abwasserentsorgung und Gewässerbelastung aus historischer Sicht am Beispiel Kiels und der Kieler Förde, Diploma thesis, CAU-Kiel, 76 pp.

Kreisel K. and Leipe T. (1989) Zum Vorkommen rezenter benthischer Foraminiferen im Greifswalder Bodden. *Wissenschaftliche Zeitschrift der Ernst-Moritz-Arndt-Universität Greifswald, mathematisch-naturwissenschaftliche Reihe* **38** (1-2), 98–104.

Kremling K., Otto, C. and Petersen, H. (1979) Spurenmetall-Untersuchungen in den Förden der Kieler Bucht. *Berichte aus dem Institut für Meereskunde an der CAU Kiel* **66**, 38 p.

References

Kucera M., and Malmgrem B.A. (1998) Logratio transformation of compositional data – a resolution of the constant sum constraint. *Marine Micropaleontology* **34**, 117-120.

Jarke J. (1961) Beobachtungen über Kalkauflösung an Schalen von Mikrofossilien in Sedimenten der westlichen Ostsee, *Deutsche Hydrographische Zeitschrift* **14**, 6-11.

Jorissen, F.J. (1999) Benthic foraminiferal microhabitats below the sediment-water interface, *in* Sen Gupta, B.K. (ed.), Modern Foraminifera: Kluwer Academic Publishers, MA, p. 161–179.

LANU (2001a) Organozinnverbindungen in Hafensedimenten und biologische Effekte, Untersuchungen in Sedimenten und an der Strandschnecke (Littorina littorea L.) in schleswig-holsteinischen Küstengewässern. Landesamt für Natur und Umwelt des Landes Schleswig-Holstein, Open Report, 55 pp. Available online at: www.lanu.landsh.de.

LANU (2001b) Deutsch-Dänische Messprogramm Flensburger Förde: Ergebnisse 1996-1997. Landesamt für Natur und Umwelt des Landes Schleswig-Holstein, Flintbek, 39 p.

LANU (2003) Sauerstoffmangel in der westlichen Ostsee im Sommer und Herbst 2002. In *Jahresbericht 2002*. Landesamt für Natur und Umwelt des Landes Schleswig-Holstein, pp. 133-139, Flintbek.

LANU (2007) Sauerstoffmangel im bodennahen Wasser der westlichen Ostsee im September 2007, MURSYS Ostsee. Landesamt für Natur und Umwelt des Landes Schleswig-Holsteins, Flintbek. Available on-line at: http://www.bsh.de/de/Meeresdaten/Beobachtungen/MURSYS-Umweltreportsystem/Mursys_031/seiten/oso27_01.jsp#september2007

Le Cadre V., Debenay J.-P., and Lesourd M. (2003) Low pH effects on *Ammonia beccarii* test deformation: implications for using test deformations as a pollution indicator. *Journal of Foraminiferal Research* **33**, 1-9.

Le Cadre, V. and Debenay J.P. (2006) Morphological and cytological responses of Ammonia (foraminifera) to copper contamination: Implication for the use of foraminifera as bioindicators of pollution. *Environmental Pollution* **143**, 304-317.

Lehmann G. and Röttger R. (1997) Techniques for the concentration of foraminifera from coastal salt meadow sediments. *Journal of Micropaleontology* **16**, 144-144.

Lehmann, G., Röttger, R., and Hohenegger, J. (2006) Life cycle variation including trimorphism in the foraminifer *Trochammina inflata* from North European salt marshes. Journal of Foraminiferal Reseach, **36**, 279-290.

Leipe T., Tauber F., Brügmann L., Irion G. und Hennings U. (1998) Schwermetallverteilung in Oberflächensedimenten der westlichen Ostsee (Arkonabecken, Mecklenburger/Lübecker Bucht and Kieler Bucht). *Meyniana* **50**, 137-154.

Leutenegger, S. (1977) Reproductive cycles of larger foraminifera and depth distributions of generations: Utrecht Micropaleontological Bulletin, **15**, 26-34.

Levander K.M. (1894) Materialen zur Kenntniss der Wasserfauna in der Umgebung von Helsingfors, mit besonderer Berücksichtung der Meeresfauna. I. Protozoa. *Acta Societatis pro Fauna et Flora Fennica* **12**, 1-115.

Linke, P. and Luze G.F. (1993) Microhabitat preferences of benthic foraminifera – a static concept or a dynamic adaptation to optimize food acquisition? *Marine Micropaleontology* **20**, 215-234.

Lutze G. (1965) Zur Foraminiferen-Fauna der Ostsee. *Meyniana* **15**, 75-142.

Lutze, G.F. (1968). Siedlungs-Strukturen rezenten Foraminiferen. *Meyniana,* **18**: 31-34.

Luze G.F. (1974) Foraminiferen der Kieler Bucht (Westliche Ostsee): 1. «Hausgartengebiet» des Sonderforschungsbereiches 95 der Universität Kiel. *Meyniana* **26,** 9-22.

Luze G.F., Mackensen A. and Wefer G. (1983) Foraminiferen der Kieler Bucht: 2. Salinitätsansprüche von *Eggerella scabra* (Williamson). *Meyniana* **35**, 55-65.

Lutze, G.F. & Altenbach, A. (1991). Technik und Signifikanz der Lebenfärbung benthischer Foraminiferen mit Bengalrot. *Geologisches Jahrbuch,* **128**: 251-265.

Mantoura, R.F.C., Dickson, A., and Riley, J.P. (1978) The complexation of metals and humic materials in natural waters: Estuarine and coastal marine sciences, **6**, 387-408.

Matthäus W. (2006) The history of investigation of salt-water inflows into the Baltic Sea – form the early beginning to recent results, *Marine Science Reports* **65**, 81 p. Available online at: http://www.io-warnemuende.de/research/mebe.html.

Meier H.E.M., Döscher R., Broman B., and Piechura J. (2004) The major Baltic inflow in January 2003 and preconditioning by smaller inflows in summer/autumn 2002: a model study. *Oceanologia* **46** (4), 557-579.

Meier,H.E.M., Feistel, R., Piechura, J., Arneborg, L., Burchard, H., Fiekas, V., Golenko, N., Kuzmina, N., Mohrholz, V., Nohr, C., Paka, V.T., Sellschopp, J., Stips, A. & Zhurbas, V. (2006). Ventilation of the Baltic Sea deep water: a brief review of present knowledge from observations and models. *Oceanologia,* **48**:133-164.

References

Mikhalevich, V.I. (1976) New data on the foraminifera of the groundwaters of Middle Asia: *International Journal of Speleology* **8**, 167-175.

Miller A.A.L., Scott D.B., and Medioli F. (1982) *Elphidium excavatum* (Terquem): ecophenotypic versus subspecific variation. *Journal of Foraminiferal Research* **12**, 116-144.

Moodley L. and Hess Ch. (1992) Tolerance of infaunal benthic foraminifera for low and high oxygen concentrations. *Biological Bulletin* **183**, 94-98.

Möbius K. (1888) Bruchstücke einer Rhizopodenfauna der Kieler Bucht. Physikalische Abhandlungen der Königlichen Akademie der Wissenschaften zu Berlin, pp. 1-31.

Murray, J.W. (1963) Ecological experiments on Foraminiferida. *Journal of the Marine Biological Association of the UK* **43**, 621-642

Murray J.W. (1989) Syndepositional dissolution of calcareous foraminifera in modern shallow-water sediments. *Marine Micropaleontology* **15**, 117-121.

Murray J.W. and Bowser S. (2000) Mortality, protoplasm decay rate, and reliability of staining techniques to recognize "living" foraminifera: a review. *Journal of Foraminiferal Research* **30**, 66-70.

Murray J.W. (2006) Ecology and application of benthic foraminifera, Cambridge University Press, Cambridge, 438 p.

Müller P.J. and Schneider R. (1993) An automated leaching method for the determination of opal in sediments and particulate matter. *Deep-Sea Research* **40**, 425-444.

Müller, G., Dominik, J., and Reuther, R. (1980) Sedimentary record of environmental pollution in the Western Baltic Sea. *Naturwissenschaften* **67**, 595-600.

Myers E.H. (1943) Life activities of Foraminifera in relation to marine ecology. American Philosophical Society, *Proceedings* **86**, 439-458.

Nausch G., Matthäus W. and Feistel R. (2003a) Hydrographical and hydrochemical conditions in the Gotland Deep are between 1992 and 2003. *Oceanologia* **45**, 557-569.

Nausch G., Feistel R., Lass H.U., Nagel K. und Siegel H. (2003b) Hydrographisch-chemische Zustandseinschätzung der Ostsee 2002. *Meereswissenschaftliche Berichte* **55**, 1-71.

Nausch G., Feistel R., Lass H.U., Nagel K. und Siegel H. (2004) Hydrographisch-chemische Zustandseinschätzung der Ostsee 2003. *Meereswissenschaftliche Berichte* **59**, 1-79.

References

Nigam, R., and Caron, D.A. (2000) Does temperature affects dimorphic reproduction in benthic foraminifera? A culture experiment on *Rosalina leei*: Current Science, **79**, 1105-1106.

Nigam, R., Saraswat, R., and Panchang, R. (2006) Application of foraminifers in ecotoxicology: retrospect, perspect and prospect: Environmental International, **32**, 273-283.

Nikulina A., Polovodova I. and Schönfeld J. (2007) Foraminiferal response to environmental changes in Kiel Fjord, SW Baltic Sea. *eEarth Discuss.*, **2**, 191-217. Available online at: http://www.electronic-earth-discuss.net/2/191/2007/eed-2-191-2007.pdf

Nikulina A., and Dullo W.-Ch., (2009). Eutrophication and heavy metal pollution in Flensburg Fjord, western Baltic Sea: a reassessment of bottom sediments after 30 years. *Marine Pollution Bulletin* **58**, 905-915.

Olsson I. (1976) Distribution and ecology of foraminiferan *Ammotium cassis* (Parker) in some Swedish estuaries. *Zoon* **4**, 137-147.

Pertillä M. (2003) Contaminants in the Baltic Sea sediments. Results of the 1993 ICES/HELCOM Sediment baseline study. *MERI – Report Series of the Finnish Institute of Marine Research*, **50**, 1-69.

Pohl C., Hennings U., Leipe T. (2005) Ostsee-Monitoring die Schwermetall-Situation in der Ostsee im Jahre 2004. *Meereswissenschaftliche Berichte* **62**, 1-36.

Polovodova, I. & Schönfeld, J. (2008). Foraminiferal test abnormalities in the western Baltic Sea; *Journal of Foraminiferal Research*, **38**: 318-336.

Rathburn A.E., Perez M.E. and Lange, C.B. (2001) Benthic-pelagic coupling in the Southern California Bight: Relationship between sinking organic material, diatoms and benthic foraminifera. *Marine Micropaleontology* **43**, 261-271.

Reeder, R.J., Schoonen, M.A.A., and Lanzirotti, A. (2006) Metal speciation and its role in bioaccessibility and bioavailability: Reviews in Mineralogy and Geochemistry, **64**, 59-113.

Reuss N., Conley D.J. and Bianchic T.S. (2005) Preservation conditions and the use of sediment pigments as a tool for recent ecological reconstruction in four Northern European estuaries. *Marine Chemistry* **95**, 3-4.

Revsbech N.P. (1989) An oxygen microelectrode with a guard cathode. *Limnology and Oceanography* **34**, 472-476.

RIIA (Royal Institute of International Affairs) (1990) Chronology and Index of the Second World

References

War 1938-1945, Westport, London, 450 p. Available online at: http://www.questiaschool.com/read/90208730?title=1944.

Rheinheimer G. (1970) Mikrobiologische und chemische Untersuchungen in der Flensburger Förde. *Berichte der Deutschen Wissenschaftlichen Kommission für Meeresforschung,* **21** (1-4), 420-429.

Rheinheimer G. (1998) Pollution in the Baltic Sea, *Naturwissenschaften* **85**, 318-329.

Rhumbler L. (1935) Rhizopoden der Kieler Bucht, gesammelt durch A. Remane, 1. Teil. *Schriften des Naturwissenschaftlichen Vereins Schleswig-Holstein* **21**, 143-194.

Rottgardt D. (1952) Mikropaläontologisch wichtige Bestandteile rezenter brackischer Sedimente an den Küsten Schleswig-Holsteins. *Meyniana* **1**, 169-228.

Rüdel H. (2003) Case study: bioavailability of tin and tin compounds. *Ecotoxicology and Environmental Safety* **56**, 180-189.

Samir A.M., and El-Din A.B. (2001) Benthic foraminiferal assemblages and morphological abnormalities as pollution proxies in two Egyptian bays. *Marine Micropaleontology* **41**, 193-127.

Saraswat R, Kurtarkar S.R., Mazumder A. and Nigam R. (2004) Foraminifers as indicators of marine pollution: a culture experiment with *Rosalina leei*. *Marine Pollution Bullein* **48**, 91-96.

Schafer C.T. (1973) Distribution of foraminifera near pollution sources in Chaleur Bay. *Water Air and Soil Pollution* **2**, 219-233.

Sharifi A.R., Croudace L.W., and Austin R.L. (1991) Benthic foraminiferids as pollution indicators in Southampton Water, southern England, United Kingdom. *Journal of Micropaleontology* **10**, 109-113.

Schiewer U. and Gocke, K. (1995) Ökologie der Bodden und Förden, In *Meereskunde der Ostsee,* 2 (ed. Rheinheimer G.), pp. 216-221. Springer Verlag, Berlin.

Schönfeld J. and Numberger L. (2007a) Seasonal dynamics and decadal changes of benthic foraminiferal assemblages in the western Baltic (NW Europe). *Journal of Micropaleontology* **26**, 47-60.

Schönfeld J. and Numberger L. (2007b) The benthic foraminiferal response to the 2004 spring bloom in the western Baltic Sea. *Marine Micropaleontology* **65**, 78-95.

Schulz F. (2000) Trendauswertung der stofflichen Belastung schleswig-holsteinischer

Fließgewässer, In *Landesamt für Natur und Umwelt Jahresbericht 1999*, pp. 59-65. LANU, Flintbek. Available online at: www.umweltdaten.landsh.de/nuis/upool/gesamt/ _jahrbe99/Trendauswertung.pdf

Schulze, F.E. (1875) Rhizopodenstudien: Archiv für mikroskopische Anatomie, **11**, 34-139.

Schwarzer K. and Themann S. (2003) Sediment distribution and geological buildup of Kiel Fjord (Western Baltic Sea). *Meyniana* **55**, 91-115.

Seiglie, G.A. (1975) Foraminifers of Guayanilla Bay and their use as environmental indicators: Revista Espanola de Micropaleontologia, **7**, 453-487.

Sen Gupta B.K., Turner R.E., and Rabalais N.N. (1996) Seasonal oxygen depletion in continental shelf waters of Louisiana: Historical record of benthic foraminifers. *Geology* **24**, 227-230.

Senocak T. (1995) Schwermetalluntersuchung an Fischen der deutschen Ostseeküste (Kliesche, *Limanda limanda;* Flunder, *Platichthys flesus;* Hering *Clupea harengus* und Dorsch, *Gadus morhua*), Berichte aus dem Institut für Meereskunde an der CAU-Kiel **270**, 177 pp.

Severin, K.P., (1990) Heterogenous trace element distribution in foraminiferal tests: what paleoceanographic resolution can be achieved? Geological Society of America, Program with Abstracts, **22**, p. 62.

Sharifi, A.R., Croudace, L.W., and Austin, R.L. (1991) Benthic foraminiferids as pollution indicators in Southampton Water, southern England, United Kingdom: Journal of Micropaleontology, **10**, 109-113.

Siegel, F.R. (2002) Environmental geochemistry of potentially toxic metals: Springer-Verlag, Berlin, 218 p.

Stouff V., Debenay J.-P. and Lesourd M. (1999a) Origin of double and multiple tests in benthic foraminifera: observations in laboratory cultures. *Marine Micropaleontology* **36**, 189-204.

Stouff V., Geslin E., Debenay J.-P. and Lesourd M. (1999b) Origin of morphological abnormalities in Ammonia (Foraminifera): studies in laboratory and natural environments. *Journal of Foraminiferal Research* **29**, 152-170.

Stouff V., Lesourd M. and Debenay J.-P. (1999c) Laboratory observations on asexual reproduction (schizogony) and ontogeny of *Ammonia tepida* with comments on the life cycle. *Journal of Foraminiferal Research* **29**, 75-84.

Ter Jung C. (1992) Beitrag zum Schwermetallgehalts-Monitoring (Zn, Cd, Hg, Cn, Ag, Pb, Cr, Ni) in

References

Miesmuscheln an der schleswig-holsteinischen Ostseeküste (1988-1989), *Berichte aus dem Institut für Meereskunde an der CAU-Kiel* **221**, 89 pp.

Tessier, A., and Campbell, P.G.C. (1987) Partioning of trace metals in sediments: relationships with bioavailability: Hydrobiologia, **149**, 43-52.

Themann S. (2002) Quartärgeologischer Aufbau und Sedimentverteilung in der Kieler Förde, Dipl. Arb.: Inst. für Geowissenschaften, CAU-Kiel, 108 p.

Tomas E., Gapotchenko T., Varekamp E.C., Mecray E.L., Buchholtz ten Brink M.R. (2000) Maps of benthic foraminiferal dustribution and environmental changes in Long Island Sound between 1940s and 1990s. In *US Geological Survey Open-File Report 00-304* (eds. V.F. Paskevich and L.J. Poppe), USGS, Woods Hole, MA. Available online at: http://pubs.usgs.gov/of/2000/of00-304/htmldocs/chap09/index.htm

Tyszka J., Topa P., and Saczka K. (2005) State-of-the-art in modelling of foraminiferal shells: searching for an emergent model. *Studia Geologica Polonica* **124**, 143-157.

Tyszka J. (2006) Morphospace of foraminiferal shells: results from the moving reference model. *Lethaia*, **39**, 1-12.

Vance D.J., Culver S.J., Corbett D.R., and Buzas M.A. (2006) Foraminifera in the Albemarle estuarine system, North Carolina: distribution and recent environmental change, *Journal of Foraminiferal Research* **36**, 15-33.

V.-Balogh K. (1988). Heavy metal pollution from a point source demonstrated by mussel (*Unio pictorium L.*) at Lake Balaton, Hungary. *Bulletin of Environmental Contamination and Toxicology* **41**, 910-914.

Wahl M. (1984) The fluffy sea anemone *Metridium senile* in periodically oxygen depleted surroundings. *Marine Biology* **81**, 81-86.

Wahl M. (1985) The recolonisation potential of *Metridium senile* in an area previously depopulated by oxygen defficiency. *Oecologia* **67**, 255-259.

Walker, D.A. (1991) Etching of the test surface of benthonic foraminifers due to ingestion by the gastropod *Littorina littorea* Linne: Canadian Journal of Earth Science, **8**, 1487-91.

Wasmund N., Pollehne F., Postel L., Siegel H. and Zettler M.L. (2005) Biologische Zustandseinschätzung der Ostsee im Jahre 2004. *Meereswissenschaftliche Berichte* **64**, pp. 1-78.

References

Wasmund N., Pollehne F., Postel L., Siegel H. and Zettler M.L. (2006) Biologische Zustandseinschätzung der Ostsee im Jahre 2005. *Meereswissenschaftliche Berichte* **69**, 1-87.

Wassman P. (1990) Relationship between primary and export production in the boreal coastal zone of the North Atlantic. *Limnology and Oceanography* **35**(2), 464-471.

Watkins, J.G. (1961) Foraminiferal ecology around the Orange County, California, ocean sewer outfall. *Micropaleontology* **7**, 199-206.

Wedepohl, K. H. (1960) Spurenanalytische Untersuchungen an Tiefseetonen aus dem Atlantik: Ein Beitrag zur Deutung der geochemischen Sonderstellung von pelagischen Tonen: Geochimica et Cosmochimica Acta, **18**, 200-231.

Wefer, G. (1976) Umwelt, Produktion und Sedimentation benthischer Foraminiferen in der westlichen Ostsee. *Reports Sonderforschungsbereich 95 Wechselwirkung Meer-Meeresboden* **14**, 1-103.

Wefer, G. and Lutze, G.F. (1976), Carbonate production by benthic foraminifera and accumulation in the western Baltic. *Limnology and Oceanography* **23**(5), 992-996.

Wennrich, V., Meng S., and Schmiedl, G. (2007) Foraminifers from Holocene sediments of two inland brackish lakes in central Germany. *Journal of Foraminiferal Research* **37**, 318-326.

Widerlund, A. & Andersson, P.S. (2006). Strontium isotopic composition of modern and Holocene mollusc shells as a paleosalinity indicator for the Baltic Sea. *Chemical geology*, **232**: 54-66.

Worm, B., and Lotze, H.K. (2006) Effects of eutrophication, grazing, and algal blooms on rocky shores: Limnology and Oceanography, **51**, 569-579.

Yanko, V., Kronfeld, A. and Flexer, A. (1994) The response of benthic foraminifera to various pollution sources: implications for pollution monitoring. *Journal of Foraminiferal Research* **24**, 1-17.

Yanko, V., Ahmad, M. and Kaminski, M. (1998) Morphological deformities of benthic foraminiferal tests in response to pollution by heavy metals: implications for pollution monitoring. *Journal of Foraminiferal Research*, **28**, 177-200.

Yanko V., Arnold A. J., Parker W.C. (1999) Effects of marine pollution on benthic Foraminifera. In *Modern Foraminifera*, (ed. B.K. Sen Gupta), pp. 217–235. Kluwer Academic Publishers, MA.

APPENDIX

APPENDIX 1

Faunal reference list of benthic foraminiferal species, considered in this thesis.

Ammonia beccarii (Linné) = *Nautilus beccarii* Linné, 1758; Schönfeld and Numberger, 2007a, p. 52, pl.1, fig.2. (Note: *Ammonia tepida*; De Noijer, 2007, p. 24, pl.1, fig. A; molecular types of *Ammonia* T1 and T2, Hayward et al., 2004, p. 256-258, pl. II-IV).

Ammotium cassis (Parker) = *Lituola cassis* Parker, 1870; Frenzel et al., 2005, p. 75, Fig. 4., no. 3.

Eggerelloides scaber (Williamson) (Note: Eggerella scabra of Lutze (1983); *Eggereloides scabrus* of Frenzel et al. (2005)).

Elphidium albiumbilicatum (Weiss) = *Nonion pauciloculum* Cushman subsp. *albiumbilicatum* Weiss, 1954; Frenzel et al., 2005, p. 73, Fig. 2., no. 10; Schönfeld and Numberger, 2007a, p. 52, pl.1, fig.4. (Note: *Elphidium asklundi* Brotzen, 1943 of Lutze (1965); *Cribroelphidium albiumbilicatum* of Frenzel (2005)).

Elphidium excavatum excavatum (Terquem) = *Polistomella excavata* Terquem, 1875, Miller et al., 1982, p. 127, pl.1, fig.11-12; Schönfeld and Numberger, 2007a, p. 52, pl.1, fig.12-13.

Elphidium excavatum clavatum (Cushman), 1930; Miller et al., 1982, p. 127, pl.1, fig.8; Schönfeld and Numberger, 2007a, p. 52, pl.1, fig.7-9.

Elphidium gerthi van Voorthuysen, 1957; Lutze, 1965, p. 159, pl. 15, fig. 45 (Note: *Cribrononion* cf. *gerthi* of author).

Elphidium gunteri Cole, 1931; Frenzel et al., 2005, p. 73, fig. 2., no. 2 (Note: *Cribroelphidium gunteri* of authors).

Elphidium incertum (Williamson) = *Polystomella umbilicatula* (Walker) var. *Incerta* Williamson, 1858; Schönfeld and Numberger, 2007a, p. 52, pl.1, fig.5-6.

Elphidium williamsoni Haynes, 1973 (Note: *Cribrononion* cf. *alvarezianum* Orbigny, 1839 of Lutze (1965)); Frenzel et al., 2005, p. 73, fig. 2., no. 8. (Note: *Cribroelphidium williamsoni* of authors).

Reophax dentaliniformis f. *regularis* Höglund, 1947; Schönfeld and Numberger, 2007a, p. 52, pl.1, fig.1.

APPENDIX 2

Appendix 2.1: Location and depth of sampling stations in the Kiel Fjord.

Stations	Date	Longitude (°E)	Latitude (°N)	Depth (m)
P0220-35.2	09.07.1996	10° 0.19'	54° 47.04'	26.3
P0220-37.2	10.07.1996	10° 15.98'	54° 27.66'	18.6
PF15-01	09.12.2005	10°10.119'	54°20.207'	11.3
PF15-02	09.12.2005	10°10.043'	54°20.266'	13.2
PF15-03	09.12.2005	10°10.486'	54°20.437'	4.6
PF15-04	09.12.2005	10°10.423'	54°20.415'	8.7
PF15-05	09.12.2005	10°10.342'	54°20.336'	13.2
PF15-06	09.12.2005	10°10.157'	54°20.324'	13.1
PF15-07	09.12.2005	10°10.473'	54°20.530'	4.6
PF15-08	09.12.2005	10°10.328'	54°20.541'	9.0
PF15-09	09.12.2005	10°10.293'	54°20.563'	12.5
PF15-10	09.12.2005	10°10.144'	54°20.557'	13.1
PF15-11	09.12.2005	10°10.435'	54°20.725'	5.0
PF15-12	09.12.2005	10°10.319'	54°20.719'	11.4
PF15-13	09.12.2005	10°10.160'	54°20.713'	13.4
PF15-14	09.12.2005	10°10.458'	54°20.888'	5.0
PF15-15	09.12.2005	10°10.360'	54°20.895'	11.2
PF15-16	08.12.2005	10°10.202'	54°20.931'	13.4
PF15-17	08.12.2005	10°12.948'	54°24.339'	4.5
PF15-18	08.12.2005	10°12.870'	54°24.325'	8.0
PF15-19	08.12.2005	10°12.699'	54°24.308'	12.2
PF15-20	08.12.2005	10°27.516'	54°27.323'	18.1
PF15-21	09.12.2005	10°09.087'	54°21.411'	10.4
PF15-22	10.02.2006	10°10.414'	54°19.781'	7.3
PF15-23	10.02.2006	10°10.251'	54°19.833'	7.2
PF15-24	10.02.2006	10°09.966'	54°19.900'	11.0
PF15-25	10.02.2006	10°09.753'	54°19.981'	13.3
PF15-26	17.02.2006	10°10.576'	54°21.048'	7.6
PF15-27	17.02.2006	10°10.501'	54°21.072'	11.6
PF15-28	17.02.2006	10°10.295'	54°21.116'	13.3
PF15-29	10.02.2006	10°09.020'	54°21.702'	11.4
PF15-30	10.02.2006	10°09.121'	54°21.575'	13.2
PF15-31	10.02.2006	10°09.155'	54°21.481'	12.6
PF15-32	10.02.2006	10°09.291'	54°21.371'	13.2
PF15-33	10.02.2006	10°09.388'	54°21.216'	13.2
PF15-34	17.02.2006	10°10.401'	54°21.820'	14.0
PF15-35	10.02.2006	10°09.599'	54°20.793'	12.5
PF15-36	10.02.2006	10°08.961'	54°19.408'	16.0
PF15-37	17.02.2006	10°11.546'	54°23.185'	16.9
PF15-38	17.02.2006	10°12.704'	54°25.233'	16.8
PF15-39	10.02.2006	10°09.356'	54°20.012'	12.8
PF15-40	10.02.2006	10°11.246'	54°23.176'	13.5
PF15-41	10.02.2006	10°10.837'	54°23.177'	12.6
PF15-42	10.02.2006	10°10.837'	54°23.270'	9.5
PF15-43	10.02.2006	10°10.856'	54°23.344'	6.2
PF15-45	17.02.2006	10°10.496'	54°21.740'	12.6
PF15-46	17.02.2006	10°10.127'	54°21.813'	14.3
PF15-47	17.02.2006	10°12.464'	54°23.528'	8.8
PF15-48	17.02.2006	10°12.288'	54°23.591'	14.9

Appendix 2 – Kiel Fjord

PF15-49	17.02.2006	10°12.429'	54°23.591'	10.5
PF15-50	17.02.2006	10°12.383'	54°23.667'	13.0
PF15-51	10.02.2006	10°09.752'	54°19.771'	15.3
PF15-52	10.02.2006	10°09.314'	54°19.519'	15.2
PF15-53	28.03.2006	10°09.933'	54°20.801'	14.8
PF15-54	28.03.2006	10°09.907'	54°21.208'	13.8
PF15-55	28.03.2006	10°09.944'	54°22.306'	14.1
PF15-56	28.03.2006	10°10.128'	54°22.620'	14.2
PF15-57	28.03.2006	10°11.557'	54°22.615'	10.9
PF15-58	04.05.2006	10°11.843'	54°25.586'	7.4
PF15-59	28.03.2006	10°11.664'	54°25.982'	17.5
PF15-60	04.05.2006	10°17.532'	54°30.315'	16.2
PF15-61	04.05.2006	10°19.063'	54°31.105'	12.5
PF15-90	05.05.2006	10°09.347'	54°19.432'	12.2

Appendix 2.2: Foraminiferal census data (percentages) of the living assemblages in the Kiel Fjord, size fraction 63-2000 μm.

Stations	Ammonia beccarii	Ammotium cassis	Elphidium albiumbilicatum	Elphidium e. clavatum	Elphidium e. excavatum	Elphidium incertum	Reophax dentaliniformis regularis	Elphiidium gerthi	Elphidium williamsoni	Elphidium guntheri	Counted specimens	Population density, ind.10cm-3	Test abnormalities, %	Species with abnormal tests*
P0220-37.2	90.0						10.0				10	3.1	no	
P0220-35.2		6.3	2.1	73.7		6.3		1.0			95	29.7	no	
PF15-01											0	-	-	
PF15-02	60.9				39.1						23	10.5	13.0	am, el ex ex
PF15-03	57.8				38.8			3.4			147	272.7	12.0	am, el ex ex
PF15-04	22.6			5.5	66.4			4.8	0.7		146	67.1	1.4	am, el ex ex
PF15-05	51.6		5.4	4.1	41.8						122	125.9	5.7	am, el ex ex
PF15-06	34.7			9.0	56.3						245	1568.0	3.8	am, el ex ex
PF15-07	25.3		2.2		72.2						79	21.0	3.8	am, el ex ex
PF15-08	27.7		2.2		56.9	4.6		10.0			130	135.4	3.1	am, el ex ex
PF15-09	48.6			2.4	49.0						255	868.1	4.7	am, el ex ex
PF15-10	51.5		3.3	28.7	18.1						171	342.0	2.4	am, el ex ex
PF15-11	57.1		3.3		36.7						49	8.2	6.0	am, el ex ex
PF15-12	72.5		2.2		25.1	1.2		0.6			346	424.5	1.4	el ex ex
PF15-13	84.1			11.6	2.9	1.4					138	447.9	2.2	am, el ex cl
PF15-14	55.3				37.4			7.3			179	703.9	3.9	am, el ex ex
PF15-15	58.0			13.0	21.0	8.0					138	581.1	1.4	am, el ex ex
PF15-16	59.9		2.2	12.6	25.3	1.5					269	1484.1	2.6	am
PF15-17	80.2		3.3		15.1			2.4			126	44.2	19.8	am
PF15-18	32.6		13.0		52.2			2.2			92	39.1	4.3	am
PF15-19	68.1		3.6	6.2	21.8	6.2		0.3			385	810.5	7.0	am, el ex ex, el inc, el ex cl, el alb
PF15-20	21.4		9.7	44.7	5.8	17.5				1.0	103	238.8	3.9	am, el ex ex
PF15-21	46.5		3.5	7.0	36.8	3.5		1.8			114	536.5	2.6	am, el ex ex, el alb

Appendix 2 – Kiel Fjord

ID											
PF15-22	53.4		7.9	37.6		1.1		189	1314.8	17.5	am, el ex ex
PF15-23	72.5		2.8	20.2		4.6		109	684.4	14.7	am, el ex ex
PF15-24	79.3		3.0	13.6		4.0		198	1413.0	11.1	am, el ex ex
PF15-25	71.0		1.0	24.5		3.5		200	1405.6	6.5	am, el ex ex, el ger
PF15-26	12.0		6.7	61.3	10.7	1.3	8.0	75	84.6	5.3	am, el ex cl, el ex ex
PF15-27	79.9		0.7	5.8	12.9		0.7		139	1263.2	4.3	am
PF15-28	68.1			6.7	22.1	3.1			163	3798.5	2.5	am
PF15-29	72.0			12.2	13.3	1.8		0.7	271	2043.6	12.9	am, el ex ex, el inc
PF15-30	46.0		1.0	15.0	30.0	3.0		3.0	1.0	100	956.6	3.0	am, el ger
PF15-31	43.5		0.8	21.0	27.4	0.8		6.5		124	561.9	3.2	am, el ex ex, el ger
PF15-32	62.8		0.7	16.0	17.4	2.4		0.7		288	995.2	4.2	am, el ex ex
PF15-33	62.4		0.0	15.1	18.8	1.4		2.3		218	1074.3	5.5	am, el ex cl, el ex ex
PF15-34	43.2		0.5		51.8	3.2		1.4		220	721.3	4.5	am, el ex ex
PF15-35	56.3		1.0	12.5	26.0	4.2				96	451.8	4.2	am
PF15-36	79.5				20.5					117	260.0	1.7	am, el exc ex
PF15-37	47.1		5.9		36.1	10.9				119	280.0	4.2	am, el ex ex
PF15-38	3.0			12.5	84.5					168	4895.1	4.2	el ex ex
PF15-39	79.4		0.5	3.4	13.7	2.0		1.0		204	2199.8	11.3	am, el ex ex
PF15-40	55.5		0.7	25.5	13.9	4.4				274	550.1	5.1	am, el ex ex, el inc
PF15-41	47.8		0.5	14.0	37.1			0.5		186	551.2		
PF15-42	51.4	0.7	3.4	12.8	29.7	1.4		0.7		148	341.9	6.8	am, el ex ex
PF15-43	34.7		2.0	1.3	45.3	8.0		8.7		150	81.1	3.3	am, el ex ex
PF15-45	70.2			6.0	22.4	1.1		0.3		352	1448.6	5.1	am
PF15-46	44.9			10.8	38.0	6.3				158	239.0	3.2	am, el exc ex
PF15-47	80.3		0.5	7.1	9.8	2.2				183	1255.4	8.7	am, el ex ex
PF15-48	62.9	0.8	0.8	4.4	27.9	1.6		1.6		251	1249.9	6.0	am, el ex ex, el ger
PF15-49	58.3		1.2	10.4	28.2	0.6		1.2		163	191.8	6.7	am
PF15-50	71.8			6.6	21.2			0.4		241	2298.9	1.2	am
PF15-51	50.8		0.3	8.8	33.1	7.0				329	2117.9	2.1	am, el ex ex, el inc
PF15-52	76.6			0.8	18.9	1.1		2.6		265	749.7	5.7	am, el ex ex, el ger
PF15-53	35.1			3.1	60.8	1.0				288	332.0	5.2	am, el exc ex
PF15-54	54.2			0.0	2.4	2.4		7.2	0.4	249	1010.2	10.4	am, el ex ex, el inc, el ger
PF15-55	52.1		1.0	10.0	25.1	4.7	0.9	5.7		211	446.5	6.6	am, el ex ex, el ger

Appendix 2 – Kiel Fjord

Sample														
PF15-56	48.8	1.2		22.8	19.1	6.2		1.9			162	122.0	3.1	am, el ex cl
PF15-57	37.8		4.4	32.8	15.7	7.6	0.3	1.5			344	1151.4	7.0	am, el ex cl, el inc, el ger
PF15-58											1			
PF15-59	3.8		2.2	54.3	32.3	6.5				1.1	186	1454.3	5.4	el ex ex, el ex cl
PF15-60	3.1		1.0	9.3	53.6	32.0		1.0			97	36.8	3.1	el inc
PF15-61	9.8		7.1	21.9	39.9	20.2		0.5	0.5		183	60.9	3.3	am, el inc, el ex ex, el ex cl
PF15-90	68.8			10.6	13.5	6.5	0.6				170	347.1	5.3	am, el inc
Mean	52.0	19.8	2.6	13.2	31.2	5.5	2.6	2.8	0.6	0.9	174	806.6	5.7	

Lutze`s samples**:

Sample									
342 (PF15-36)	15.4			80.8	3.8			26	**50.0**
341 (PF15-35)		12.5		68.8	18.8			16	**22.5**
340 (PF15-34)			3.5	7.0	57.9	14.0	15.8	1.8	57	**11.3**
239 (PF15-38)	10.4	3.8		46.2	24.5	13.2		0.9	106	**11.0**
Mean	12.9	6.6		26.6	58.0	12.5	15.8	1.3	51	**23.7**

*am indicates the species *Ammonia beccari*, el ex ex – *Elphidium excavatum excavatum*, el ex cl – *E. excavatum clavatum*, el inc – *E. incertum*, el alb – *E. albiumbilicatum*, el ger – *E. gerthi*.

** The **bold** numbers indicate data taken from the Lutze`s manuscript (Lutze, 1965)

Appendix 2 – Kiel Fjord

Appendix 2.3: The living and dead percentages of foraminifera in Kiel Fjord in 1963 (Lutze, 1965) and 2006.

Sample	1963*			2006		
	Living ind. %	Dead ind. %	Living/ Dead ratio	Living ind. %	Dead ind. %	Living/ Dead ratio
342 (PF15-36)	15	85	0.2	47	53	0.9
341 (PF15-35)	15	85	0.2	24	76	0.3
340 (PF15-34)	0	100	0.0	40	60	0.7
238 (PF15-37)	20	80	0.3	38	62	0.6
239 (PF15-38)	10	90	0.1	76	24	3.2
mean	12	88	0.2	45	55	1.1

* The living and dead foraminifera percentages in 1963 are taken from Lutze (1965)

Appendix 2 – Kiel Fjord

Appendix 2.4: Correlation matrix of environmental parameters and foraminiferal data from Kiel Fjord. Am bec = *Ammonia beccarii*; El ex = *Elphidium excavatum excavatum*; El ex cl = *E. excavatum clavatum*; El gert = *E. gerthi*; El wil = *E. williamsoni*; El alb = *E. albiumbilicatum*; El gunt = *E. gunteri*; A cas = *Ammotium cassis*; R dent = *Reophax dentaliniformis*. The **bold** numbers indicate significant correlations (Student t-test. a = 0.05).

n = 40	Sand, %	SiO₂, %	C$_{org.}$, %	TN, %	C:N ratio	Chl a, ng/g	Cu, ppm	Zn, ppm	Sn, ppm	Pb, ppm	Pop den, ind/10cm³	Abnorm tests, %	A:E Index	Am bec, %	El ex, %	El ex cl, %	El gert, %	El wil, %	El alb, %	El inc, %	El gunt, %	A cas, %	R dent, %
Sand, %	1.000																						
SiO₂, %	**-0.736**	1.000																					
C$_{org.}$, %	**-0.797**	**0.825**	1.000																				
TN, %	**-0.498**	**0.472**	**0.511**	1.000																			
C:N ratio	0.226	**-0.351**	-0.075	**-0.662**	1.000																		
Chl a, ng/g	**-0.378**	**0.605**	**0.402**	**0.280**	**-0.372**	1.000																	
Cu, ppm	**-0.580**	**0.458**	**0.737**	**0.312**	0.128	0.031	1.000																
Zn, ppm	**-0.614**	**0.499**	**0.782**	**0.319**	0.131	0.071	**0.946**	1.000															
Sn, ppm	**-0.384**	**0.414**	**0.584**	**0.424**	-0.080	0.121	**0.318**	**0.286**	1.000														
Pb, ppm	**-0.635**	**0.508**	**0.754**	**0.350**	0.127	0.042	**0.928**	**0.950**	**0.318**	1.000													
Pop den, ind/10cm³	-0.226	0.190	0.213	0.231	-0.252	0.263	-0.122	-0.056	**0.341**	-0.043	1.000												
Abnorm tests, %	0.147	0.125	0.077	0.027	-0.067	0.108	-0.054	0.051	0.140	-0.023	0.039	1.000											
A:E index	-0.244	**0.341**	**0.349**	0.165	-0.067	0.104	0.078	0.150	**0.332**	0.125	0.029	0.163	1.000										
Am bec, %	**-0.296**	**0.406**	**0.337**	0.263	-0.098	-0.098	**0.330**	**0.310**	**0.295**	**0.347**	-0.042	**0.307**	**0.442**	1.000									
El ex, %	0.151	**-0.406**	-0.228	-0.207	0.156	-0.148	-0.018	-0.029	-0.165	-0.096	0.148	-0.148	**-0.394**	**-0.521**	1.000								
El ex cl, %	-0.257	**0.303**	0.157	0.098	-0.170	**0.622**	-0.185	-0.125	-0.043	-0.089	0.171	-0.103	-0.029	**-0.338**	-0.269	1.000							
El gert, %	-0.034	-0.033	0.111	0.114	0.043	-0.117	0.165	0.171	0.150	0.109	-0.185	0.177	0.062	-0.095	0.255	**-0.296**	1.000						
El wil, %	0.014	-0.058	0.039	-0.046	0.181	-0.116	0.312	0.311	-0.112	0.284	-0.129	0.025	-0.093	-0.210	0.296	-0.043	0.196	1.000					
El alb, %	**0.359**	**-0.280**	**-0.416**	-0.255	-0.173	-0.061	**-0.364**	**-0.365**	**-0.351**	**-0.372**	**-0.293**	0.026	-0.178	-0.246	0.027	0.216	-0.043	-0.106	1.000				
El inc, %	0.073	-0.046	-0.178	-0.112	-0.093	0.070	**-0.340**	**-0.339**	-0.075	-0.277	-0.135	-0.085	-0.032	-0.257	-0.223	**0.534**	0.020	-0.108	**0.419**	1.000			
El gunt, %	-0.141	0.221	0.107	0.117	-0.261	0.608	-0.199	-0.130	0.039	-0.186	0.032	0.054	0.078	**-0.394**	-0.128	**0.713**	-0.048	-0.043	0.285	**0.539**	1.000		
A cas, %	0.254	-0.271	-0.152	-0.151	0.566	-0.172	-0.175	-0.199	-0.001	-0.131	-0.119	-0.082	-0.028	-0.020	-0.103	0.183	0.014	-0.026	-0.106	0.156	-0.043	1.000	
R dent, %	-0.090	0.056	0.042	0.033	-0.103	0.175	-0.034	-0.028	0.113	0.092	0.063	0.033	-0.033	-0.100	-0.133	**0.314**	-0.011	-0.026	0.148	0.215	-0.043	-0.026	1.000

APPENDIX 3

Appendix 3.1: Census data on living foraminiferal assemblages from two profiles done by Exon (1972) in southern Gelting Bay (Profile 10525) and outer Flensburg Fjord (Profile 10528). Percentages of every species were recalculated from archive data, stored at Institute of Geosciences (University of Kiel). Note the absence of living individuals at depth 22 m in southern Gelting Bay. IOL indicates here the inner organic lining.

Area	Station	Latitude (N)	Longitude (E)	Water depth (m)	Sediment type	Ammonia beccarii	Ammoscalaria runiana	Ammotium cassis	Elphidium excavatum	Elphidium incertum	Miliammina fusca	Hippocrepina	Reophax dentaliformis	Absolute number of living ind. (N)	Remarks
Profile 10525	525-1	54° 46.625'	9° 52.750'	8	Muddy-sand	2.7	1.8	83.9	0.9					111	IOL
	525-2	54° 46.876'	9° 52.750'	10	Muddy-sand	4		76		16.0	4			25	Corrosion. IOL
	525-3	54° 47.060'	9° 52.850'	12	Sandy-mud			89.5		8.80	1.8			57	Corrosion. IOL
	525-4	54° 47.060'	9° 52.750'	14	Sandy-mud	6.1		57.6		36.4				33	Solution. IOL
	525-5	54° 47.120'	9° 52.500'	16	Sandy-mud			10	10.0	80				20	Some solution
	525-6	54° 47.370'	9° 52.375'	18	Sandy-mud			13.3		80				15	A few IOL
	525-7	54° 47.685'	9° 52.000'	20	Sandy-mud			9.1	18.2	54.5		6.7	18.2	11	Some solution. IOL
	525-8	54° 48.377'	9° 51.375'	22	Mud										Some solution
Profile 10528	528-10	54° 49.815'	9° 53.500'	11	Sand					82.6	17.4			23	Much corrosion. IOL
	528-12	54° 49.750'	9° 53.490'	12	Sand	1.3		96.2		2.6				78	IOL
	528-11	54° 49.813'	9° 53.490'	16	Sand					62.5	37.5			8	Corrosion weaked?
	528-7	54° 49.813'	9° 53.688'	18	Muddy-sand			33.7		54.7			11.6	95	Some IOL
	528-6	54° 49.813'	9° 53.625'	20	Sandy-mud			8.3	25.0	58.3			8.3	12	*E. incertum* with sand, a few IOL
	528-5	54° 49.820'	9° 53.688'	22	Sandy-mud			12.5	12.5	50			25	8	
	528-4	54° 49.950'	9° 53.750'	24	Sandy-mud				11.1	88.9				9	*E. incertum* with agglut. sand grains
	528-3	54° 50.100'	9° 54.100'	26	Mud				33.3	66.7				3	
	528-2	54° 50.563'	9° 55.185'	28	Mud				28.9	71.1				38	Solution

Appendix 3 – Flensburg Fjord

Appendix 3.2: Foraminiferal census data of stations, sampled in 2006 in Flensburg Fjord.

Sample	Ammonia beccarii	Ammotium cassis	Elphidium albiumbilicatum	Elphidium excavatum excavatum	Elphidium excavatum clavatum	Elphidium gerthi	Elphidium gunteri	Elphidium incertum	Elphidium williamsoni	Reophax dentaliniformis	Species richness (S)	Absolute number of ind. (N)	Population density, ind./10cm^3
PF16-01	29.8		1.1	64.9	3.2			1.1			5	94	49.5
PF16-02	11.1		6.9	59.7	22.2						4	72	16.4
PF16-03	66.0		3.4	8.1	17.9			4.7			5	235	353.7
PF16-04			10.2	34.2		0.5		58.3			4	187	324.2
PF16-05	2.1		4.2	2.1	1.4			91.0		6.3	6	144	157.6
PF16-06			0.9					98.7		0.9	3	230	105.3
PF16-07			1.5	57.3	8.4	0.8		32.1			5	131	78.9
PF16-08	11.8	0.7	3.9	8.5	8.5			66.7			6	153	126.7
PF16-09													
PF16-10	4.7		30.2	53.5	11.6						4	43	21.2
PF16-11	28.4		4.2	47.4	16.8			3.2			5	190	131.6
PF16-12	32.5	0.8	20.8	22.5	4.2			20.8			6	120	286.3
PF16-13	9.0		1.6	23.3	15.9			49.2		1.1	6	189	77.2
PF16-14	5.2		1.3	58.7	19.4			15.5			5	155	69.6
PF16-15	11.1			6.6	26.1			55.8			5	226	199.4
PF16-16	48.2		12.9	25.9				12.2		0.7	5	139	98.3
PF16-17	37.0		6.7	25.2	8.9	0.7		20.7	0.7		7	135	192.1
PF16-18	46.5	0.6	12.1	5.1	5.7			29.3		0.6	7	157	347.9
PF16-19	17.9		5.4	14.3	5.4			57.1			5	56	28.3
PF16-20	95.8		2.1	2.1							3	142	234.3
PF16-21	70.3		7.0	4.7	2.3			14.8		0.8	6	128	121.2
PF16-22	18.4	0.7	26.5	29.4	13.2			11.0		0.7	7	136	758.7
PF16-23	16.1	0.8	8.1	4.0	2.4			68.5			6	124	82.5
PF16-24	47.5		49.4	1.9				1.2			4	162	253.7
PF16-25	24.7	1.3	56.0	16.7				1.3			5	150	229.6
PF16-26	94.4									5.6	2	18	11.3
PF16-27	5.5	0.8	3.1	53.5	36.2			0.8			6	127	142.2
PF16-28	6.0	0.5	1.1	51.9	35.0		1.1	4.4			5	183	937.7
PF16-29	3.5	0.8	4.7	22.9	56.6	1.2		10.1			8	258	1510.4
PF16-30	5.2		2.1	55.4	29.3			4.2			5	287	3130.2
PF16-31	10.4		20.2					68.1		1.2	4	163	200.2
PF16-32	7.8	0.9	3.5	18.3	4.3			65.2			6	115	101.5

APPENDIX 4

Appendix 4.1: Absolute amount of the different test abnormality modes observed in foraminiferal species of the Kiel Fjord.

Foraminiferal species experienced test abnormalities	Aberrant chamber shape	Twisted and distorted chamber arrangement	Additional chamber	Reduced size of chambers	Over developed chambers	Protuberances	Multiple appertures	Wrong coiling	Poor development of the last whorl	Development of two different whorls	Twinning	Lack of sculpture	Spiroconvex tests	Compressed tests	Bulla-like chamber at the umbilicum	Twisting of the entire test	Non-developed test	Complex form	Total number of abnormal tests	Species proportion of all abnormalities (%)	Average proportion in all samples (%)
Ammonia beccarii	103	29	12	120	76	26	2	22	1	1	7	1	75	45	2	14	2	5	543	73	66
Elphidium excavatum excavatum	24	7	5	43	1	6	1	3	1	2	2			20		3	2	1	121	16	31
Elphidium excavatum clavatum	2			14	2	1		1		3				5					28	4	28
Elphidium gerthi	3		3	11							1			3					22	3	17
Elphidium incertum	3	1		6	3	2	1	3		2	2			2					24	3	29
Elphidium albiumbilicatum			1	1	3														5	1	12

Appendix 4 – Test abnormalities

Appendix 4.2: Absolute amount of the different test abnormality modes observed in foraminiferal species of the Flensburg Fjord.

Foraminiferal species experienced test abnormalities	Aberrant chamber shape	Twisted and distorted chamber arrangement	Additional chamber	Reduced size of chambers	Overdeveloped chambers	Protuberances	Multiple apertures	Wrong coiling	Poor development of the last whorl	Development of two different whorls	Irregular keel	Twinning	Spiroconvex tests	Compressed tests	Bulla-like chamber at the umbilicum	Twisting of the entire test	Complex form	Total number of deformed tests	Species proportion of all abnormalities (%)	Average proportion in all samples (%)
Ammonia beccarii	23	1		31	13	7		11		1		4	37	7	2	6	1	144	40	56
Elphidium excavatum excavatum	9		1	22	9	2		2	1		1	2		7				56	15	36
Elphidium excavatum clavatum	9		1	13	1	1	1	2		1	1			5		2		36	10	27
Elphidium gerthi	1																	1	0.3	7
Elphidium incertum	26	1	8	29	19	6		10		2		5		9	1	1	1	117	32	44
Elphidium albiumbilicatum	2		1	5				1						1				10	3	14

Appendix 4 – Test abnormalities

Appendix 4.3: Correlations revealed between the abnormality modes and environmental parameters in Kiel and Flensburg fjords. The significance-test for a linear correlation at normal distribution of data was performed according to Aßmann and others (2007). Note that a type I error $\alpha=0{,}05$. Bold numbers indicate the correlations, which passed a significance-test i.e., have T-value higher than $t_{n-2,\,1-\alpha}$.

Area	Abnormality mode	Environmental parameter	Correlation coefficient (r)	Significance-test, (T)	Quantile, $(t_{n-2,\,1-\alpha})$
Kiel Fjord	Additional chamber	Sn	-0.516 (n=10)	-1.706	1.812
	Wrong coiling	C_{org}	**-0.608** (n=9)	-2.026	1.895
	-	N_{total}	**-0.610** (n=9)	-2.037	1.895
	-	Chlorophyll a	**-0.824** (n=9)	-3.848	1.895
Flensburg Fjord	Additional chamber	Cu	**0.902** (n=5)	3.619	2.35
	-	Zn	**0.865** (n=5)	2.986	2.35
	-	Sn	**0.850** (n=5)	2.795	2.35
	-	Pb	**0.861** (n=5)	2.932	2.35
	-	N_{total}	-0.626 (n=6)	1.605	2.132
	Twisting of entire test	Cu	0.627 (n=6)	1.610	2.132
	-	Zn	0.564 (n=6)	1.366	2.132
	-	Sn	0.615 (n=6)	1.560	2.132
	-	Pb	0.720 (n=6)	2.075	2.132

APPENDIX 5

PLATES

Appendix 5 – PLATES

PLATE 1

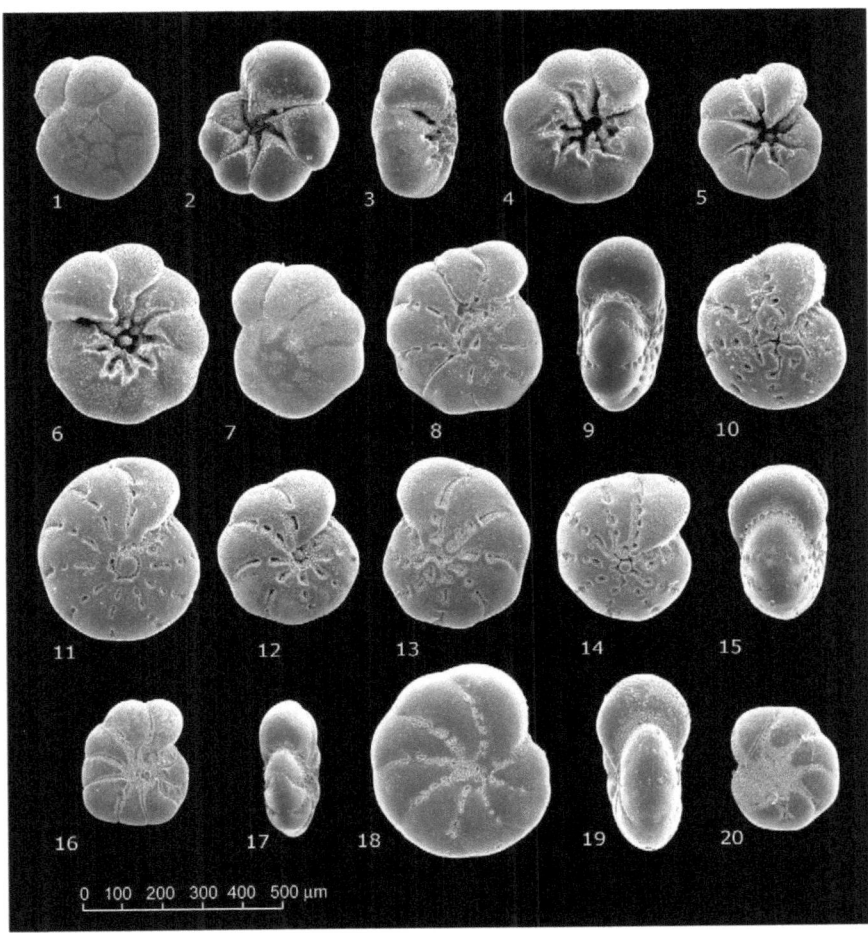

XVI

PLATE 1

Some species of living benthic foraminifera from Kiel Fjord.

FIGURE 1-7. *Ammonia beccarii.*

    Poorly ornamented spiral sides (1,7); apertural view (3);

    umbilical view (2, 4-6).

FIGURE 8-13. *Elphidium excavatum excavatum.*

    Spiral view (8,11-13); apertural view (9).

FIGURE 14-15. *Elphidium excavatum clavatum,* spiral and apertural view.

FIGURE 16-17. *Elphidium gerthi,* spiral and apertural view.

FIGURE 18-19. *Elphidium incertum,* spiral and apertural view.

FIGURE 20. *Elphidium albiumbilicatum*, spiral view.

PLATE 2

PLATE 2

Some species of living benthic foraminifera from Flensburg Fjord.

FIGURE 1-4. *Ammonia beccarii*: spiral (1. 2) and umbilical (3. 4) views.

FIGURE 5-7. *Elphidium excavatum clavatum*: spiral view (5, 6); apertural view (7) and detailed view of mineralogical projections (7a).

FIGURE 8-11. *Elphidium excavatum excavatum*: spiral (8. 10) and apertural (9. 11) views.

FIGURE 12-15. *Elphidium incertum*: spiral view (12); apertural view (13) and remains of the cyst at the test surface (14, 15).

FIGURE 16. *Reophax dentaliniformis*.

FIGURE 17-19. *Elphidium albiumbilicatum*: spiral view (16. 18); apertural view (17) and detailed view of mineralogical projections at the test surface (18a).

FIGURE 20. *Ammotium cassis*.

FIGURE 21-22. *Ammonia beccarii* with frequent test perforation (Gelting Bay; uncoated SEM images)

Appendix 5 – PLATES

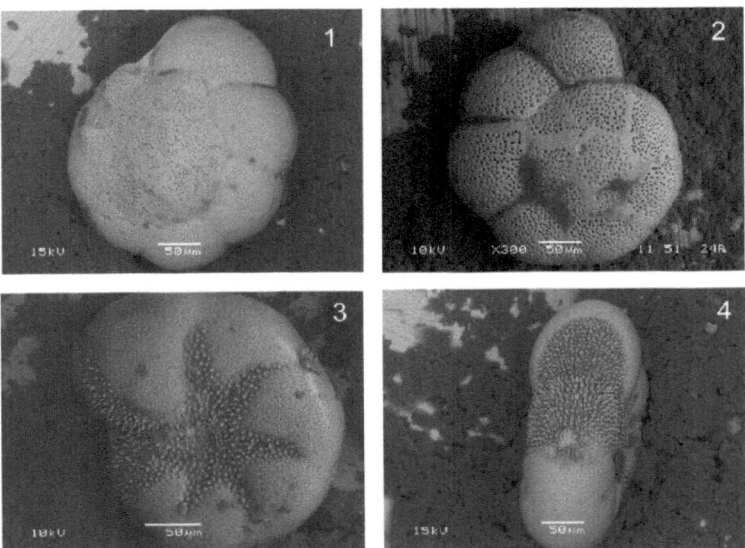

PLATE 3. Uncoated SEM images of foraminifera from Flensburg Fjord. **1-2.** Different porosity of *Ammonia beccarii* tests. Note the smaller pores of a test from sample PF16-25 (1) as compared to higher porosity of *Ammonia* test taken from PF16-20 (2). **3-4.** *Elphidium albiumbilicatum*: spiral (3) and apertural (4) planes. Note the pustules in the umbilical and apertural area.

Appendix 5 – PLATES

PLATE 4

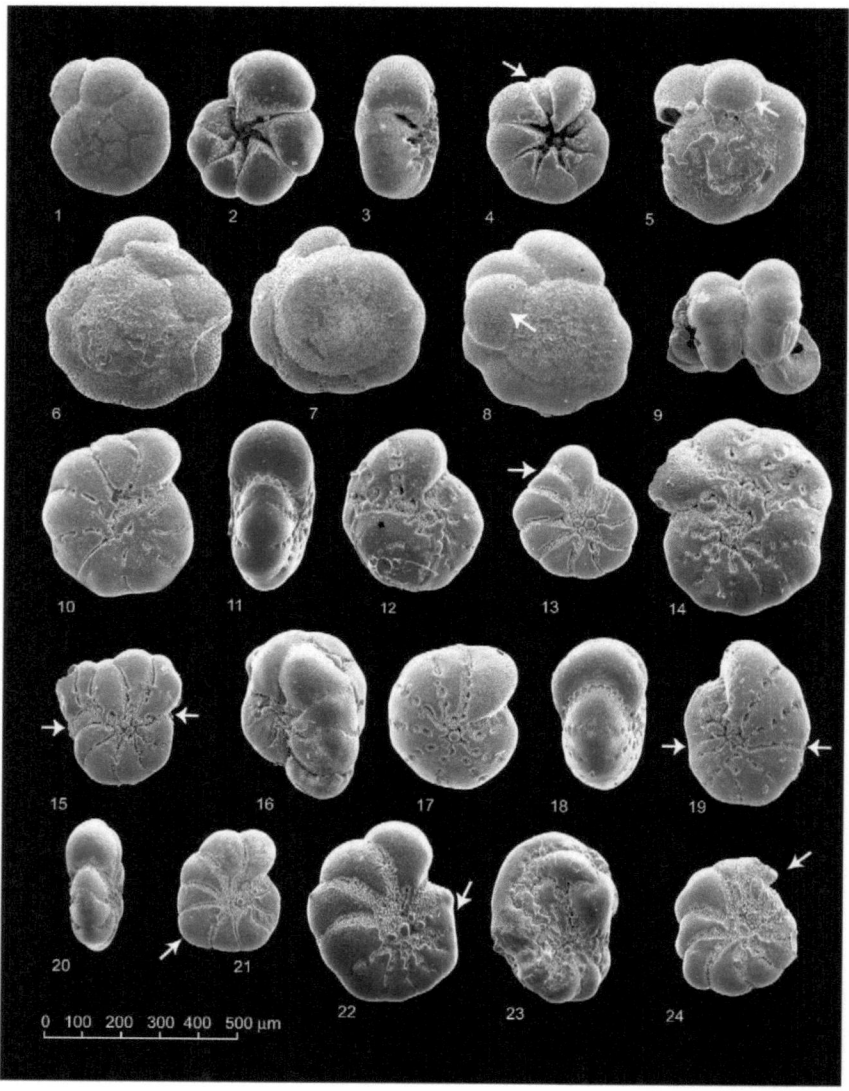

XXI

Appendix 5 – PLATES

PLATE 4
Normal and abnormal specimens encountered in Kiel Fjord.

FIGURE 1-9. *Ammonia beccarii.*

1-3. Spiral, umbilical and apertural views of normal specimen. **4.** Abnormal test with reduced chamber size (arrow). **5.** Additional chamber at the spiral side of the test. **5.** Note the regeneration scars at the basement of an additional chamber (arrow). **6.** Aberrant shape of the last chambers. **7.** Abnormal spiroconvex test with a distinctly high spiral side. **8.** Additional chamber at the spiral side of the test. **9.** The double test (twins) showing the fusion of two specimens of the same size by spiral sides.

FIGURE 10-16. *Elphidium excavatum excavatum.*

10-11. Spiral and apertural views of normal specimen. **12.** Abnormal test with a lack of sculpture at the spiral side. **13.** Reduced chamber size (arrow). **14.** Overdeveloped chambers of the last whorl. **15.** Double apertures (arrows). **16.** Abnormal test exhibiting the development of several different whorls.

FIGURE 17-19. *Elphidium excavatum clavatum.*

17-18. Spiral and apertural view of normal specimen.

19. Compressed test (arrows).

FIGURE 20-24. *Elphidium gerthi.*

20. Apertural view of normal specimen. **21.** Abnormal test possessing a slightly overdeveloped chamber (arrow). **22.** Reduced chamber size (arrow). **23.** Abnormal specimen with a complex form of abnormality creating difficulties with taxonomical identification of it as *E. gerthi*. **24.** Poor development of the last whorl (arrow).

Appendix 5 – PLATES

PLATE 5

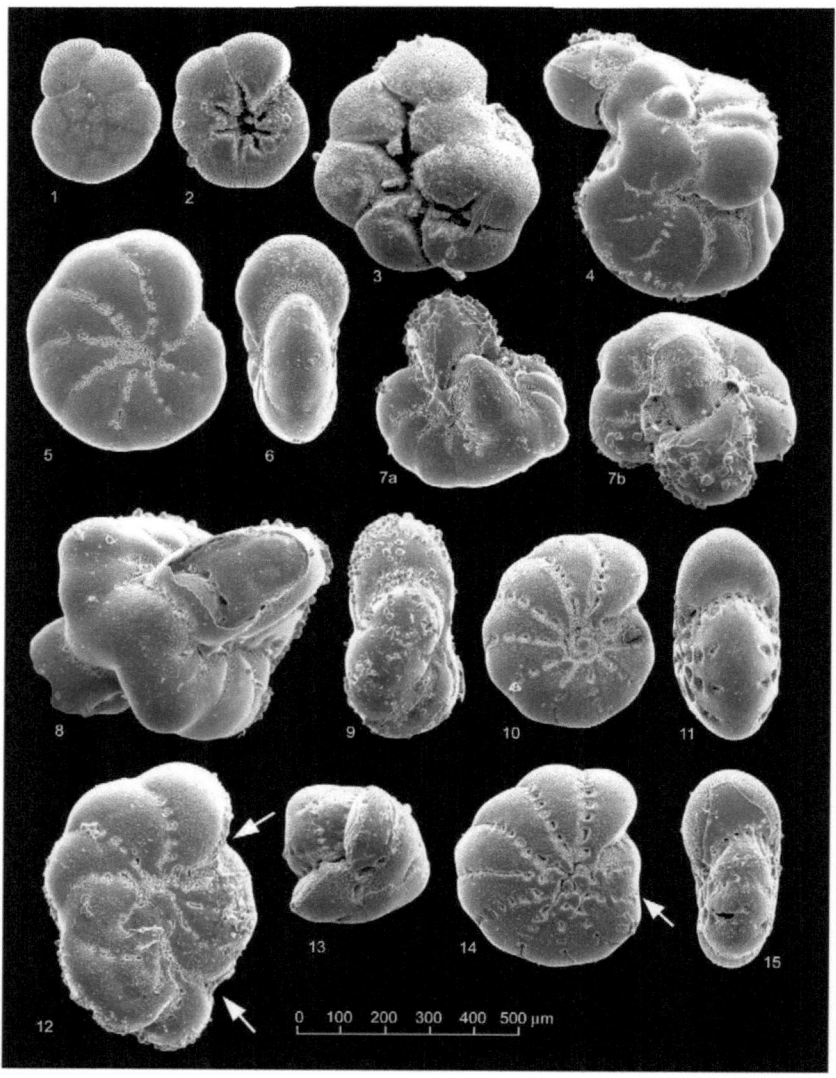

Appendix 5 – PLATES

PLATE 5.

Normal and abnormal specimen observed in Flensburg Fjord.

FIGURE 1-3. *Ammonia beccarii.*

    Spiral and umbilical views of normal specimen(1-2).

    Compressed test, umbillical view (3).

FIGURE 4-9; 12. *Elphidium incertum.*

    Double test (4)

    Spiral and apertural views of normal specimen (5-6).

    Development of two different whorls (7ab; 8)

    Twisting of entire test (9)

    Double apertures (12)

FIGURE 10-11; 14-15. *Elphidium excavatum excavatum.*

    Spiral and apertural views of normal specimen (10-11)

    Reduced chamber size (14)

    Twisting of entire test (15)

Appendix 5 – PLATES

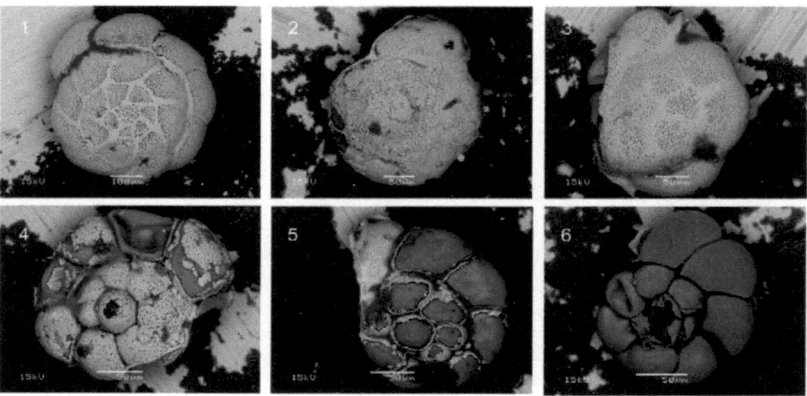

PLATE 6. Different stages of test dissolution observed in tests *A. beccarii* from the Gelting Bay, Flensburg Fjord: normal test (1); loss of first chambers, which are thinner (2-3), note the initial dissolution process taken place at the spiral side of the test (2) and visible organic lining at the place of a dissolved first chamber (3); dissolution of the following chambers (4); heavily dissolved test with interlocular walls remained or so called star-shaped form (5); elastic organic lining folded in places (6). Scale bar = 50 μm. Note, that images are taken with JSM-6460LV SEM without coating.

Appendix 5 – PLATES

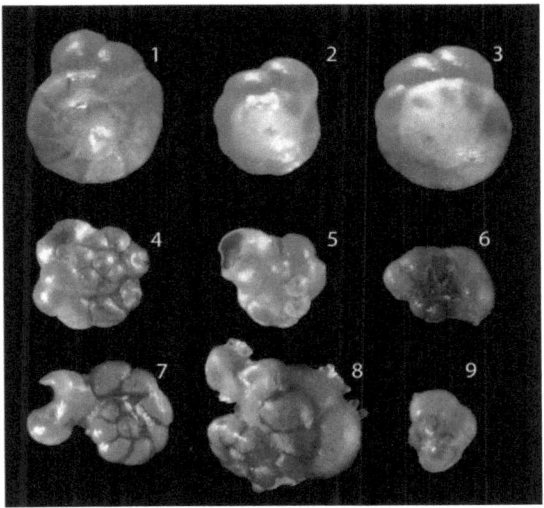

PLATE 7. Light microscopic images of *A. beccarii* tests from the Gelting Bay, Flensburg Fjord. **1**. Normal specimen, spiral view. **2-3**. Opaque tests, as the first stage of test dissolution. **4-9**. Double tests showing the smaller specimen exhibiting disruptions in the coiling plane (**7-8**) and firmly attached to the partially or fully destroyed bigger foraminifer.

i want morebooks!

Buy your books fast and straightforward online - at one of world's fastest growing online book stores! Environmentally sound due to Print-on-Demand technologies.

Buy your books online at
www.get-morebooks.com

Kaufen Sie Ihre Bücher schnell und unkompliziert online – auf einer der am schnellsten wachsenden Buchhandelsplattformen weltweit! Dank Print-On-Demand umwelt- und ressourcenschonend produziert.

Bücher schneller online kaufen
www.morebooks.de

VDM Verlagsservicegesellschaft mbH
Heinrich-Böcking-Str. 6-8   Telefon: +49 681 3720 174   info@vdm-vsg.de
D - 66121 Saarbrücken   Telefax: +49 681 3720 1749   www.vdm-vsg.de

Printed by Books on Demand GmbH, Norderstedt / Germany